Ecodocumentaries

Rayson K. Alex • S. Susan Deborah
Editors

Ecodocumentaries

Critical Essays

palgrave
macmillan

Editors
Rayson K. Alex
Birla Institute of Technology and Sciences, Pilani, Goa, India

S. Susan Deborah
M. E. S. College of Arts & Commerce, Goa, India

ISBN 978-1-137-56223-4   ISBN 978-1-137-56224-1 (eBook)
DOI 10.1057/978-1-137-56224-1

Library of Congress Control Number: 2016957775

Printed on acid-free paper

This Palgrave Macmillan imprint is published by Springer Nature
The registered company is Macmillan Publishers Ltd.
The registered company address is: The Campus, 4 Crinan Street, London, N1 9XW, United Kingdom

To

All the ecodocumentary filmmakers who support tiNai Ecofilm Festival

Foreword: Packaging Concerns

"They didn't fear their demise, they repackaged it—it can be enjoyed as video games, as TV shows, books, movies. The entire world wholeheartedly embraced the apocalypse and sprinted toward it with gleeful abandon." Thus spoke Governor David Nix in the film *Tomorrowland*.

We are they, and we are having a heyday with all things eco. Ecocinema, ecomedia, ecodocumentaries, ecoperformances, ecocriticism: it is the best of times, and it is the worst of times to be the natural environment. There is intense interest—popular and academic—in all things eco. There is hope, and there is despair. There are utopic visions of sustainable "futuristic" futures (clean, glossy, digital, silicon), and there are dystopic views of imagined horrors to continue (dirt, *WALL-E*, and the gray of *The Road*). There is entertainment, and there is fact (the difference between the two increasingly not so great). Clever scriptwriters give words to gifted actors, who then serenade us with speeches about the state of things, undoubtedly with more charisma and passion than most academics or activists can ever hope to achieve, not to mention an infinitely broader reach. There are, to be sure, really good results from all of this. "Clean technologies" are developing very quickly indeed. So too, however, are the subsequent new problems that often attend these new technologies. Novel things that look truly great on the surface sometimes turn out to be truly horrendous. Take the digital revolution, for instance. Cloud storage and the Internet are wonderful and indispensable in my line of work, but there is a sobering reality about sustainability here that most of us perhaps would wish just weren't true: one of these is that by 2009, "the server farms that allow the internet to operate and that provide cloud-based digital computing had surpassed the airline industry in terms of the amount of carbon dioxide released into the earth's atmosphere" (Rust, Monani, and Cubitt "Introduction," 3). Who would have thought? And who ever thinks about the actual physical waste of the actual physical equipment? One of the documentaries covered in this collection (deftly and thoroughly by Başak Ağın Dönmez and tangentially by Chia-ju Chang) is about the mega-electronics dump in Accra, Ghana. These two things alone (the carbon dioxide usage and the burial grounds of electronics) go beyond simply shocking; they are virtually paralyzing. Real eye-openers. And opening eyes is what this collection is all about.

It is clear that "one of the central ways we shape our relationship to other animals, our place on Earth, and the social structures that arise from these understandings is through media and culture," as John Parham has eloquently explained in *Green Media and Popular Culture: An Introduction* (1).[1] Indeed, it is the centrality of media in industrialized nations to understandings of self and the world that has resulted in so many books recently addressing and exposing both humanity's spectacularly visible ecocaust and the slower aspects of this violence toward nature. These books attest to the growing importance of scholarship that urges sustained ecocritical and other eco-inflected analyses of what has come to be termed ecomedia.[2]

Two books—each fresh and wide-ranging—by Stephen Rust, Salma Molani, and Stephen Cubitt have been and continue to be vital in this new area of study: *Ecocinema Theory and Practice* (2013) is arguably the first collection of essays about eco-inflected cinema; *Ecomedia: Key Issues* (2016) is another first, massive in scope, giving more attention than ever before to the topic and ranging in discussion from thematic issues to matters of production, from divergences to convergences, from frames to flows. Another foundational text, without which *Ecodocumentaries: Critical Essays* simply would not have been possible, is Sean Cubitt's 2005 *EcoMedia*, a book that boldly states upfront that "we have no better place to look than the popular media for representations of popular knowledge" (1). Infused with biophilic and ecophobic ethics, knowledge about our natural environment is both represented in and produced by popular media.

Pat Brereton's *Environmental Ethics and Film* (2016) builds on Cubitt's important work and analyzes the medium of film and how it has been, is being, and can be used to reimagine ethical values. Also influenced by Cubitt's work, *Culture and Media: Ecocritical Explorations* (2014)—by

Rayson K. Alex, S. Susan Deborah (the editors of the current volume), and Sachindev P.S.—expands the discussions and puts India center stage in a study of a richly diverse and provocative body of ecomedia. Using Nirmal Selvamony's "*tinai* criticism," Alex identifies ecocriticism in India as "a local and global concept…an indigenous way of looking at the land and its people from natural, cultural, and supernatural perspectives" (3) and seeks "a holistic understanding of place, people, and culture" in this book, "a major portion of [which]…is devoted to essays on cinema" (4).

In light of the tremendous energy and work appearing on eco-inflected media, then, *Ecodocumentaries: Critical Essays* is a timely collection. It does what good scholarship in the environmental humanities should do. It digs deeply, it unearths, it questions, and—perhaps most importantly—it motivates. It joins and extends a conversation about media and environment, and about the role documentaries play in educating and inspiring. It brings together some of the best thinkers in this new field of scholarship—a field that is deeply challenging and filled with ambivalences.

One of the more prominent challenges of the field, as Rob Nixon has famously observed in his remarkable *Slow Violence and the Environmentalism of the Poor*, "is representational: how to devise arresting stories, images, and symbols adequate to the pervasive but elusive violence of delayed effects" (2). A key insight of John Parham, too, is that often "green messages sit uncomfortably alongside dominant (non-green) ideologies and/or anthropocentric thinking" (xx). Indeed, the challenges for media are strong, and the urge to include things that are known to sell well seems irresistible—hence, the jarring presence of sexist configurations and images in so many mainstream eco-inflected movies.

It has long been a research interest of mine to know what happens when, as John Parham so nicely words it, media "adapt[s] itself to and speak[s] in the modes and language of the dominant culture" (xvii). These modes and this language are imbued not only with the patriarchal values of sexism and heterosexism (both profitable for patriarchies) but also with the values of ecophobia (also profitable for patriarchies). Ecophobia prominently colludes with sexism any time we hear some unseemly comment about "Mother Nature" (sometimes even described as a cunning "bitch").[3]

One of the main things that many of the films covered in *Ecodocumentaries: Critical Essays* address is ecophobia, and although the word does not appear much in the following pages, many of the contributors to this volume address the concept in varying ways. Pat Brereton hints at it without actually naming it in his discussion about the E.O. Wilson's "controversial theory of 'biophilia.'" Chia-ju Chang more directly raises the matter of ecophobia, again without actually naming it but rather by describing it—in contrast to biophilia (which she does mention)—as "the dark and 'inconvenient' side of the earth's story."

The packaging of ecophobia (and sexism, heterosexism, racism, among other things) uncomfortably alongside green issues, however, is not the only danger ecodocumentaries face, not the only thing that threatens the effectiveness of ecodocumentaries, and not the only challenge ecodocumentaries must confront. I have noted elsewhere (see Estok, "Ecomedia and Ecophobia") that there are several reasons why so much of ecomedia has had limited effects on pushing people to change their behaviors, a crippled capacity to halt or slow the warming of our atmosphere, and these challenges must be addressed: (1) ecomedia not only sometimes perpetuate the ecophobic ethics that are so central to the problem in the first place; ecomedia are embedded in a period in which our continuous partial attention runs hand in hand with our compassion fatigue; (2) ecomedia dilutes the material to such a degree that important abstract concepts are blurred, thus preventing thinking people from seeing key connections; and (3) ecomedia has become entertainment, and the blurring of virtual and actual worlds makes a lot of the actual news simply another form of entertainment. Speaking "in the modes and language of the dominant culture" indeed does bring the issues to a wide audience, does make knowledge more accessible, does theoretically make a difference; yet, the tensions and contradictions of ecomedia, delivering, as it does, comments about nature being a bitch and daring people to dance[4] are manifold.

Tensions and difficult balances fill the pages of this book. The very topic implies, as Pietari Kääpä in this collection observes, a "difficulty of balancing human stories with environmental rhetoric." In the films Kääpä discusses, "even as they evoke environmentalist arguments, they also conceptualise nature as property." There is no question, as Pat Brereton also argues in this volume, that we "need much more ecodocumentaries… that speak[…] to and hopefully help[…] in the process of gently changing behaviour patterns with regards to food production alongside other related environmental concerns, before it is too late." Yet, while it certainly seems counterintuitive to argue against such an idea (and I don't want to do so here), we must also ask, first, if pummeling ourselves with data is going to do the trick and, second, whether we should address questions about information overload and numbness, about how, as Naomi Klein has recently put it, "each massive disaster seems to inspire less horror, fewer telethons" (53). Compassion fatigue increases with every news report, every additional documentary, and every fresh environmental horror story. To be clear, some of this stuff, if I may quote from Divya Anand's chapter, "is at times over-awing for the viewer." This is one of the challenges for ecodocumentaries: to provide numbing material without numbing the viewer.

Documentaries that move people and are in some ways jarring perhaps still remain the best bet. In her discussion of *In God's Land*, Salma Monani describes a kind of jarring affect, a "visual and aural affect" of an animated film that she describes as "discombobulating," the film itself being not easy to watch. It is about Special Economic Zones (SEZ) in India and the conflicts surrounding them. While the film is not anti-SEZ, it does, nevertheless, capture the sense of "the simmering conflict that surrounds the SEZ" and of how it seems to offer instant solutions of "development" for long-term problems of poverty and disenfranchisement, solutions that themselves will have "long-term ecological aftermaths."

All of the contributors to this volume are, of course, in principle concerned about the "long-term ecological aftermaths" of human activities as they are represented in diverse documentaries, but many of the chapters also share a vital concern with how documentaries package the seemingly opposing topics of the local and the global, the individual and the community, the self and the other. In his discussion of the documentary *The Queen of the Trees*, for instance, Nirmal Selvamony examines the ontology of the fig tree and stresses that although the individual is important and undeniable, community membership and belonging are also ontological realities. We ignore these at our own risk. Selvamony argues about the impossibility of isolating the individual (and of the foolishness in trying to do so) and about the importance of recognizing that the individual is always a part of a larger living body. John Duvall maintains that the sense of connectedness is of global importance, and this matter takes center stage in his discussion of the documentary entitled *Elemental*. The film documents the work of three activists in three very different parts of the world: an Indian government official taking personal action to clean up the Ganges River, a First Nations Canadian mother campaigning against the Alberta Tar Sands project, and an Australian entrepreneur/inventor trying to get investors to put money into biomimicry research and development. Duvall sees the importance of linking the local and the global, and notes that "what is ultimately so impressive about *Elemental* is the way it embraces universal spiritual, philosophical, and ecological themes, while keeping a sharp focus on individual people in the context of their families, cultures, and personal struggles." Sreejith Varma R. and Swarnalatha Rangarajan, in their discussion of Kuttan Vayali's *Bhagavathy aattu*, make a strong case for the importance of local rituals in environmental matters, of deep interrelationships between the ecocultural and the spiritual. They analyze "'Theyyam,' the folk-ritual pageant of North Malabar of Kerala, a south-western coastal state of India" and show that "the Theyyam dance is a fine example of embedded ecology since its co-ordinates of ritual and myth connects the cosmos with the materiality of the earth and interweaves the themes of fertility and prosperity for the community and its future generation." Finally, Rayson K. Alex, Samuel Moses Srinivas Kuntum, and Selvaraj Susan Deborah expose the moral carelessness of colonial bodies and their practices in their description of the British introduction of the invasive species known locally in India as cheemaikaruvel (*Prosopis juliflora*). The moral carelessness of the British has resulted in enormous environmental and social damage. In discussing their own documentary (entitled *Thorny Land: Invasion of Cheemakaruvel*), Alex, Kuntum, and Selvaraj offer a kind of supplement to their film in order to express the things they "feel have not been adequately expressed in the film." It is a fitting close to a fascinatingly diverse package of material, a collection of essays that reveals tensions, ambivalences, and the sometimes thorniness of the issues we confront.

This book says a lot about the work that ecodocumentaries do, about what gets packaged and what doesn't, about what some of the problems are and where some of the answers hide, about what are clear-cut issues and what are thorny ones, about where we've been and where we might be going. This is a hopeful book, both regional and global, both very timely and very necessary, a book that makes links and challenges the reader to pursue further connections. Most of all, this is a book that dares us to unpack issues and to generate changes.

Notes

1 This echoes Pat Brereton's comment a year earlier in *Environmental Ethics and Film* that "for most people the mass media are the primary way in which they acquire ethical attitudes, especially within contemporary culture" (2).

2 Stephen Rust defines "Ecomedia studies...as a historically situated, ideologically motivated, and ethically informed approach to the intersections, of media, society, *and* the environment" (87, *italics* in original).

3 Describing the natural environment as a "bitch" obviously involves not only sexism and ecophobia but also speciesism. A recent Brad Pitt movie entitled *World War Z*, for instance, has a doctor ranting about nature in the following manner:
"Mother Nature is a serial killer. No one's better. More creative. Like all serial killers, she can't help the urge to want to get caught. What good are all those brilliant crimes if no one takes the credit? Now the hard part—while you spend a decade in school—is seeing the crumbs for the clues there. Sometimes the thing you thought was the most brutal aspect of the virus turns out to be the chink in its armor. And she loves disguising her weaknesses as strengths. She's a bitch."

4 Alvin Duvernay, describing Hurricane Katrina's destruction of his home, states in *The Age of Stupid*:
You stare Mother Nature in the eye. Usually, she's fairly benign. Then she comes along, methodically, ruthlessly. And then she stands toe-to-toe with you and dares you. *Dares* you: "Go ahead and get your best equipment out. Go ahead. Do it. Let's dance."

References

Alex, Rayson K. "Introduction." *Culture and Media: Ecocritical Explorations.* Eds. Rayson K. Alex, S. Susan Deborah, and Sachindev P.S. Newcastle upon Tyne: Cambridge Scholars, 2014. 1–9. Print.

Brereton, Pat. *Environmental Ethics and Film.* New York: Routledge, 2015. Print.

Cubitt, Sean. *EcoMedia.* Amsterdam: Rodopi, 2005. Print.

Estok, Simon C. "Ecomedia and Ecophobia." *Neohelicon: Acta comparationis litterarum universarum* 43.1: 127-45. Print.

Klein, Naomi. *This changes everything: Capitalism vs the climate.* Toronto: Vintage, 2015. Print.

Nixon, Rob. *Slow Violence and the Environmentalism of the Poor.* Cambridge: Harvard UP, 2011. Print.

Parham, John. *Green Media and Popular Culture: An Introduction.* Basingstoke: Palgrave Macmillan, 2015. Print.

Rust, Stephen, Salma Monani, and Sean Cubitt. "Introduction: Ecologies of Media." *Ecomedia: Key Issues.* Eds. Stephen Rust, Salma Monani, and Sean Cubitt. New York: Routledge, 2015. 2–10. Print.

Rust, Stephen, Salma Monani, and Sean Cubitt, eds. *Ecocinema Theory and Practice*. New York: Routledge, 2013. Print.

Rust, Stephen. "Overview: Flow—An Ecocritical Perspective on Broadcast Media." *Ecomedia: Key Issues*. (Eds. Stephen Rust, Salma Monani, and Sean Cubitt. New York: Routledge, 2015. 85–98. Print.

The Age of Stupid. Dir. Franny Armstrong. Spanner Films (Dogwoof Pictures), 2009. Film.

Tomorrowland. Dir. Brad Bird. Walt Disney, 2015. Film.

World War Z. Dir. Marc Forster. Paramount Pictures, 2013. Film.

Simon C. Estok
estok@skku.edu
Department of English Language and
Literature College of Liberal Arts,
Sungkyunkwan University,
Seoul, South Korea

Acknowledgements

A book is a labour of love, an investment of time, and the collaborative project of many individuals. We are therefore greatly indebted to many friends, colleagues, mentors, and editors who have accompanied us along this journey. First, we would like to acknowledge our gratitude to ASLE-USA (Association for the Study of Literature and Environment) Media Subvention Grant, which has enabled this project to prosper and which was responsible for sowing the seed to publish a volume on the documentaries screened at the 2014 *tiNai* Ecofilm Festival (TEFF 2014). We sincerely acknowledge Prof. Chia-ju Chang and Prof. Ursula Heise, who mentored us in the initial process of finding a publisher and encouraged us to progress with the idea of bringing out a volume of edited essays. Our deepest gratitude goes to all the contributors for graciously supporting us by submitting their essays for publication. This volume would not have been possible without the support and encouragement of the Birla Institute of Technology and Science (BITS), Goa, which allowed us to hold TEFF 2014 on campus. We are grateful to the head of the Department of Humanities and Social Sciences, Professor Meenakshi Raman, the then directors of the Institute, Professor K.E. Raman, and the erstwhile director, Late Professor Sanjeev K. Agarwal, and the current director, Professor Sasikumar Punnekkat for having believed in us and our conviction to hold an exclusively ecological film festival on campus and for supporting us. We also gratefully acknowledge all the filmmakers who willingly sent us their films for screening: Mr. Mathieu Roy, Mr. Pankaj Rishi Kumar, Mr. Kunal Vohra, Mr. Mark Deeble and Ms. Victoria Stone, Mr. Sandeep Kr. Singh, and Ms. Sunanda Bhat. Without your films, TEFF

2014 would not have become a reality. We would also like to thank the members of the jury of TEFF 2014, Prof. K.P. Jayshankar, Prof. Anjali Monteiro and Prof. Salma Monani, the festival advisors, and all the festival participants and well-wishers for their support and encouragement.

There are certain people in our lives whose presence alone has been an invaluable addition to our endeavours—our mentors. We are deeply indebted to our teacher, mentor, and guide, Dr. Nirmal Selvamony, who has always been with us in our academic as well as our personal journeys, constantly motivating, encouraging, and aiding us whenever any clouds of doubt descended. We would also like to express our thanks to Professor Simon C. Estok and Professor Salma Monani for showing us the light at the end of the tunnel. We owe them much indeed! We thank our proofreader, Ms. Penelope Iremonger, for her hawk-eyed precision and great eye for detail. We are thankful to our editor, Ms. Lina Aboujieb, for her guidance and support and for her immediate replies to our mails. We are indebted to Mr. Solano Da Silva, Dr. Anoop George, Dr. Reena Cheruvalath, Dr. Gyan Prakash, Dr. Geetha K.A., Ms. Pragyan Barik, Dr. Geetha B., Dr. Shalini Upadhyay, Professor Alito Sequiera, and Dr. Aaron Lobo, our friends and colleagues from BITS-Goa and other prestigious institutions, who have been our pillars of strength, providing us with much-needed words of encouragement whenever necessary, and also for assisting us with the logistics of organizing a festival.

Last but definitely not least, we are eternally grateful to our families for supporting and encouraging our projects and academic endeavours throughout the years.

Rayson K. Alex
S. Susan Deborah

Tamil Transliteration Key

Consonants	
k	க்
c	ச்
T	ட்
ι	த்
p	ப்
R	ற்
g	ங்
j	ஞ்
N	ண்
n	ந்
m	ம்
n	ன்
y	ய்
r	ர்
l	ல்
v	வ்
z	ழ்
L	ள்

Vowels	
a	அ
aa	ஆ
i	இ
ii	ஈ
u	உ
uu	ஊ
e	எ
ee	ஏ
ai	ஐ
o	ஒ
oo	ஓ
au	ஔ

This Transliteration System was developed by Nirmal Selvamony (for *Indian Journal of Ecocriticism* in 2008) by suitably modifying the systems found in *Index of puRanaanuuRu* (1962) and *Grammar of akanaanuuRu with Index* (1972) by V.I. Subramonian, and *Tirukkural in Ancient Scripts* (1980) by Gift Siromoney, S. Govindarajau and M. Chandrasekaran.

Contents

List of Figures

Introduction: The Ethics of Relationships in Social Documentaries

Rayson K. Alex and *S. Susan Deborah*

The acknowledgement of the existence of disparate ecologies and ecological experiences and relationships, between humans, between humans and other organisms, and between humans and the land, water, air, and earth, converge in one single point—ethics. Like ecocinema studies, other areas within ecocriticism consider ethics to be a major concern. In their "Introduction" to *International Perspectives in Feminist Ecocriticism*, Gaard, Estok, and Oppermann clearly consider ethics as a crucial concern in feminist ecocriticism:

> Feminist ecocriticism, then, fosters an ontological understanding of ethics, raising ethical awareness about bodily natures, and promoting an ecological-feminist discourse of hope and change. In a word, feminist ecocriticism opens up new ethical pathways to contest the sexist, racist, speciesist, ecophobic, classist, nationalist and homophobic discourses of "nature," which have perpetuated gendered dualities and bodily boundaries. (Gaard et al. 2)

R.K. Alex (✉)
Department of Humanities and Social Sciences, Birla Institute of Technology and Science, Pilani, K. K. Birla Goa Campus, Zuarinagar, Goa 403726, India

S.S. Deborah
Department of English, M. E. S. College of Arts & Commerce, Zuarinagar, Goa 403726, India

R.K. Alex, S.S. Deborah (eds.), *Ecodocumentaries*,
DOI 10.1057/978-1-137-56224-1_1

Clare Palmer claims that ethics is one of the three major concerns in environmental philosophy, the others being epistemology and justice, which, of course, have their own ethical implications (187). The core of environmental politics is "sustainability ethics", as proposed by Peter Dauvergne (466). Dauvergne writes, "…an ethic of sustainability sets the stage for more analytical and eventually, political work" (466). Sustainability ethics should go hand in hand with environmental ethics, he claims. Scott Slovic considers an "ethical framework" to be the foundation for ecocritical enquiry: "Without an ethical foundation, how would we know when and where to intervene in the world's problems? The very fact that we're talking about ethics and activism in the context of a form of literary scholarship indicates what is particularly poignant about this field for many of its practitioners" (Oppermann 462).

Most discussions on the ecological humanities seem to border around an ethical framework. Scholars in ecocinema studies define ecocinema in terms of genre (Scott MacDonald, Willoquet-Maricondi), labelling ecocinema as activist documentaries, art-house cinema, and documentaries with clear environmental concerns. It is with great excitement that an environmental studies student at Gettysburg College, USA, exclaimed, "Wow! That means every and any film can be analyzed ecocritically" (Rust and Monani 1). Salma Monani and Stephen Rust, in their "Introduction" to *Ecocinema Theory and Practice*, invite a theoretical attention to the term "ecocinema," rather than looking at it as a genre of film. According to them, the scope of "ecocinema studies is not simply limited to films with explicit messages of environmental consciousness, but investigates the breadth of cinema from Hollywood corporate productions and independent avant-garde films to the expanding media sites in which producers, consumers, and texts interact" (ibid.). One of the reasons that ecocinema should be seen as a theory is that "all films present productive ecocritical exploration and careful analysis can unearth engaging and intriguing perspectives on cinema's various relationships with the world around us" (Rust and Monani 3). As mentioned earlier, the study of relationships is based on an ethical framework. The environmental ethics framework is spelt out explicitly in the book *Environmental Ethics and Film* (2015) by Pat Brereton. Brereton applies Carolyn Merchant's five assumptions of environmental ethics derived from the different kinds of relationships between entities (Merchant 76–78):

1. "Everything is connected to everything else."
   Here, connection would mean relationship. The ethics of the relationship is specific to this famous dictum.

2. "The whole is greater than the sum of its parts."
   What is "the sum of its parts"? Although an extension of the first point, the specific cultural and scientific interspecies and intraspecies connections, for functional or cultural reasons, have ethical value. However, these relationships should not be seen in isolation but as a whole, which creates a larger (than the smaller parts) world with larger ethical concerns.
3. "Meaning is context dependent."
   The context would refer to land, water, air, and mythical or historical interconnections.
4. "Process has primacy over parts."
   The way in which one organism relates with a living or non-living entity has ethical implications.
5. "Humans and non-human nature are one."
   Any possibility of a dichotomy (even in an ideological sense) is unethical.

The material ecocritical phenomena of "network of agencies" (Iovino and Oppermann 1) suggest a just process of human agency with non-human agencies. In "Introduction: Ecologies of Media," Stephen Rust, Salma Monani, and Sean Cubitt argue that ecomedia study is "the complex work of deciphering which forms of media...facilitate ecological discussion" on "how we should live." In a nutshell, ecomedia studies imply "making ethical judgments" (Rust, Monani, and Cubitt 6).

Ecodocumentaries: Critical Essays explores the various dimensions of ethics in the medium of cinema. These dimensions of ethics are embedded as "entanglements that intertwine nature and culture, science and the humanities, the knower and the known" (Alaimo 188) and "human entanglement with other humans, with the non-human world, and with contemporary and historical power dynamics that influence representations of these relationships" (White 145). Endorsed by Rust, Monani, and Cubitt in their recent volume *Ecomedia: Key Issues in Environment and Sustainability* (2015), James Farrell's idea of "commons sense" is of relevance in this context, being: "The common sense of the twenty-first century in which 'everybody knows' that human life depends on other lives in the biosphere—and the health of the biosphere itself" (256). Jason Taylor's seven-minute environmental film, *A Commons Sense* (2014), borrows Farrell's phrase to showcase the ethnographic activities of the ecologist and seed conservationist, Dr. Debal Deb, India's Rice Warrior, as he is called. Working closely with farmers across Indian villages, he has cultivated approximately 920 rice varieties in two and a half acres of forested land in the Niyamgiri hills, Odisha. The scientists' data of the probability of over 100,000 varieties of rice in India originally comes from the common sense—the farmers' sense and everybody's knowledge—that there are different varieties of rice in each locality. The documentaries analysed in the current volume both implicitly and explicitly deal with the ethics of the entanglements of common sense.

The ten chapters in this book present some important socio-environmental issues—food justice, postcolonial environmentalism, ecological imperialism, posthumanism, transcorporeality, animal justice, progress and development, river ecology and conservation, the politics of the land, and electronic waste. The chapters are divided into three parts, which are not tight compartments but "enmeshed" with each other.

Part One is entitled "Land, Food, Ethics." Salma Monani's chapter, "*In God's Land*: Cinematic Affect, Animation, and the Perceptual Dilemmas of Slow Violence," analyses an Indian film *In God's Land*, set in Tamil Nadu. Monani suggests that "*In God's Land*'s blend of animation and live action illuminates what Banu Subramanian has called 'archaic modernity'—the pervasive way religion and science and technology meld in modern India (Subramanian)," leading to the destructive nuances of slow violence. Aligning herself with the goat herders, who might soon be displaced from their ancestral land, Monani questions the various ethical aspects of the livelihood, nationalism, environmental politics, and capitalistic perspectives of development. Transnational in its methodology, Pietari Kääpä's chapter, "Transnational Perspectives on Land Ethics: *Elemental* and *Not My Land*," ecocritically analyses two films, a US-produced film *Elemental* (2012) and an Estonian film *Not My Land*. Kääpä uses the framework of Aldo Leopold's land ethic to characterise "the sorts of complex evaluation of humanity's fraught relationship with the biotic environment as it emerges in these films." Kääpä explores some ecocritical concerns, such as the "dynamics of ecosystemic relationality," the "dexterity of the ecocosmopolitan approach," and the ethics of the film context, to highlight the "shared environmental responsibility" of the viewers. Pat Brereton's chapter, "Communal Indian Farming and Food Ecology: A Reading of *Timbaktu*," focuses on a short Indian documentary and is "a persuasive cautionary tale concerning the over-use of pesticides and the general abuses of factory farming methods, while endorsing organic communal modes of farming production." Brereton identifies the farmers as "nature citizens" and brings the reader's attention to his/her environmental responsibility in supporting the farmers and their sacred relationship with the land.

Part Two is entitled "Development, Waste, Ethics." Başak Ağın Dönmez's chapter, "Ecological Imperialism in the Age of the Posthuman: David Fedele's *E-Wasteland*," reveals the neoimperialistic politics of the developed nations and questions the ethics of dumping e-wastes in third-world countries. In the chapter "Wasted Humans and Garbage Animals: Deadly Transcorporeality and Documentary Activism," Chia-ju Chang analyses several documentary films from various cultural contexts. She argues that the civilisation–garbage divide is a false conception and that the discourse on the "intersection between waste and animals contributes to current waste studies and serves as a point of departure to launch a new integrated field of 'animal-garbage' studies." Divya Anand, in her chapter "*Surviving Progress*, Modernity and Making Sense of the Crisis in Nature," establishes that the "historical understanding of progress as economic, social, and cultural development is undeniably tied to the way in which humankind has engaged with nature throughout time." While ecopolitically analysing the film *Surviving Progress*, Anand argues that any claim to protect and preserve nature or "socially develop" specific cultural communities can lead to a tension in reclaiming nature, "from small communities to the nation-state to international corporations as stakeholders."

Part Three, entitled "Tree, River, Ethics," includes four chapters. Nirmal Selvamony argues that the ontology of the fig tree, in the environmental documentary *The Queen of Trees*, is the continuation of the fig wasp and all the other organisms connected with the tree. In his chapter "What is "A" Fig?" he maintains that the tree is not an individual but a community and, thereby, claims that, like other organisms, plants are "capable of sensing the ethicality of the consciousness of their patients and reacting accordingly." John Duvall, in his chapter "*Elemental* as Existential and Mythic Parable," looks at the film *Elemental* through the lens of spiritual symbolism, literary myth, and existentialist philosophy. Sreejith Varma and Swarnalatha Rangarajan's chapter, "The Possibilities of a River and a Ritual Dance: An Ecoethnographic Analysis of Kuttan Vayali's *Bhagavathy aattu*," argues that the ethnographic documentary film *Bhagavathy aattu* brings to mind "the limits of metaphysics by emphasising our rootedness in the all-embracing gestalts of environment, culture, and other site-specific factors." They claim that "ontological agency is always constituted by the ecological." In their chapter, "Eco-exoticism in *Thorny Land: Invasion of Cheemakaruvel*," Rayson K. Alex, K. Samuel Moses Srinivas, and S. Susan Deborah argue that the ethics of ecological exoticism depend on the functions and activities of the subject itself.

All the documentaries under discussion in the current volume were screened for public viewing in the tiNai Ecofilm Festival (TEFF) 2014, held at the Birla Institute of Technology and Science Pilani, K.K. Birla Goa Campus, Goa, Western India. In accordance with the objective of the festival, we invited scholars in the field of ecocriticism and media studies to contribute chapters to a proposed volume, critiquing one or more films curated at TEFF. The diverse nationalities of both the contributors and the films, and the specific issues discussed in the films, bring forth an entanglement of ecological perspectives, all in one platform. This interesting experimental project brings together stories and scholars from Ireland, Turkey, the USA, the UK, India, China, Estonia, and Canada. A unique feature of the volume is the inter- and intra-ethno-eco-cultural dialogue between films and scholars of different nationalities—an Irish scholar critiquing *Timbaktu*, with an Indian environmental context; a Turkish scholar analysing *E-Wasteland*, presenting an African environmental context; an American-Chinese scholar engaging with Indian and Chinese films; an Indian scholar discussing the story of a fig tree in Africa; and so forth.

As ecodocumentaries continue to entertain, inspire, engage, educate, create awareness, and initiate political activism amongst people, we are sure this venture will intellectually stimulate ecocinema and ecocriticism scholars in the years to come. We hope this volume will inspire other interested scholars in India and other third-world countries to collaborate in creating further scholarly works.

References

Alaimo, Stacy. "Feminist Science Studies and Ecocriticism: Aesthetics and Entanglement in the Deep Sea." Ed. Greg Garrard. *The Oxford Handbook of Ecocriticism*. Oxford; New York: Oxford UP, 2014. Print.

Farrell, James. *The Nature of College: How a New Understanding of Campus Life can Change the World*. Minneapolis: Milkweed, 2010. Print.

Gaard, Greta, Simon C. Estok, and Serpil Oppermann, eds. "Introduction." *International Perspectives in Feminist Ecocriticism*. New York; London: Routledge, 2013. Print.

Iovino, Serenella, and Serpil Oppermann, eds. "Introduction: Stories Come to Matter." *Material Ecocriticism*. Bloomington and Indianapolis, IN: Indiana UP, 2014 Print.

MacDonald, Scott. "Toward an Eco-cinema." *Interdisciplinary Studies in Literature and Environment* 11.2" (Summer 2004) 107-32.

Merchant, Carolyn. *Radical Ecology: The Search for a Living World*. New York: Routledge, 1992. Print.

Monani, Salma, and Stephen Rust, eds. "Introduction: Cuts to Dissolves – Defining and Situating Ecocinema Studies." *Ecocinema Theory and Practice*. New York: Routledge, 2013. Print.

Palmer, Clare. "Introduction to Environmental Philosophy: Ethics, Epistemology, Justice." Ed. Ricardo Rozzi et al. *Linking Ecology and Ethics for a Changing World: Values, Philosophy, and Action*. Heidelberg; New York; London: Springer, 2013. Print.

Rust, Stephen, Salma Monani, and Sean Cubitt, eds. "Introduction: Ecologies of Media." *Ecomedia Studies: Key Issues in Environment and Sustainability*. London; New York: Routledge, 2016. Print.

Slovic, Scott, Swarnalatha Rangarajan, and Vidya Sarveswaran, eds. "Introduction." *Ecocriticism of the Global South*. Lanham, MD: Lexington Books, 2015. Print.

White, Laura. "Re-Imagining the Human: Ecofeminism, Affect and Post-Colonial Narration." Ed. Greta Gaard, Simon C. Estok, and Serpil Oppermann. *International Perspectives in Feminist Ecocriticism*. New York; London: Routledge, 2013. Print.

Willoquet-Maricondi, Paula, ed. *Framing the World: Explorations in Ecocriticism and Film*. Charlottesville and London: University of Virginia Press, 2010. Print.

PART I

Land, Food, Ethics

In God's Land: Cinematic Affect, Animation, and the Perceptual Dilemmas of Slow Violence

Salma Monani

For his 2009 documentary, *Seeds of Dissent*, Pankaj Rishi Kumar accompanied the Indian environmentalist Dr. Anil Joshi and his team on a bicycle journey in winter 2008. The "Agri-Cycle Yatra" intended to raise awareness about India's farmers, ran 3800 km from India's southern tip in Kanyakumari, Kerala, to its northern regions in the Himalayas. In the trip's early stages, the crew stopped in the Tirunelveli district of southern Tamil Nadu. Though its southern border flanks the Bay of Bengal, most of Tirunelveli's geography resembles the barren, rocky, dry lands of the Deccan region.

As Kumar put his documentarian eye to work, he noticed a large sign in the semi-arid, rural landscape announcing that a tract of land equivalent to 2500 acres (approximately 10 km^2, or the area of a small township) had been set aside as a Special Economic Zone (SEZ). Designed to encourage economic development, SEZs are controversial entities. Forwarded by a neoliberal government agenda, SEZs across the country are lauded for their corporate friendliness in a country where bureaucracy stifles economic progress, even as they are faulted for social and environmental injustices (Levien; Anwar; Kannan).

S. Monani (✉)
Department of Environmental Studies, Gettysburg College, 300 N. Washington Street, Gettysburg, PA 17325, USA

R.K. Alex, S.S. Deborah (eds.), *Ecodocumentaries*,
DOI 10.1057/978-1-137-56224-1_2

Kumar's gut reaction to the SEZ was curiosity. Why here, in this "god-forsaken landscape?" (personal communication).[1] Kumar decided to leave the cycle team for a few days and investigate. One of the first things Kumar learned was that the local Vanamamalai temple (which worships a version of the Hindu god, Lord Vishnu) had sold the 2500 acres to the federal government for the SEZ allotment. Following up on this lead, Kumar interviewed a local priest, who explained that the temple was pleased to help benefit the government as well as the people of the area.

As Kumar describes, he was personally amazed at the quiet reallocation of so much land—"wow, 2500 acres in India, and no conflict?" (personal communication). At the same time, he was fascinated by what the development of this land would mean for the primarily rural, semi-arid region and its inhabitants. Kumar decided to document the changes to the area as the SEZ developed. It was only on a later visit that Kumar learned that the "SEZ wasn't as clean as it first appeared" (personal communication).

The resulting documentary production, *In God's Land* (hereafter *IGL*) aka *Kadavulin Nilathil* (2012), has screened at various venues, from the Indian cable television station NDTV, to independent film festivals such as tiNai, to academic screenings in India's universities, as well as in classrooms in the USA. At 72 minutes, it captures a sense of the simmering conflict that surrounds the SEZ and delves into a long history of tenuous existence for Tirunelveli's poorest inhabitants—its subsistence farmers and goat herders.

In describing the film, Kumar is categorical in stating that it is not "anti-SEZ." Instead, Kumar wished to take up a bigger cinematic challenge. In my interview with him, he explains:

It is very easy to land in a conflict zone, like an anti-dam or an anti-nuclear event. There are two parties; somebody doing something and it is a power game so the conflict is apparent. In this situation, on the face of it, nothing is happening. What happens on a day-to-day basis for the villagers is not visible. For me, that is much more challenging: how can I capture this scene of violence and this madness? (Personal communication)

In effect, Kumar's challenge falls within the realm of representing what postcolonial literary scholar Rob Nixon calls "slow violence." As Nixon, Nixon writes that such violence is out of sight and attritional; it also does not lend itself to easy representation, "never materializing in one spectacular, explosive, cinematic scene" (14). Instead, as Nixon suggests, slow violence is much like Johan Galtung's concept of structural violence, which draws attention to complex, deep-seated, systemic forms of violence. What distinguishes slow violence is its particular attention to the "temporalities of place" (18). It is interested in how time generates a distinct socio-ecological disadvantage for the economically poor and disenfranchised: "What does it mean for people declared disposable by some 'new' economy to find themselves existing out of place in place, as, against the odds, they seek to slow the ecological assaults on inhabitable possibility?" (19). Violence towards the poor is thus not simply of one moment—like the development of the SEZ—but of long, attritional injustices that exacerbate such moments and leave communities living with long-term ecological aftermaths.

In this chapter, I am interested in how a documentary venture like *IGL* can capture such long-term and seemingly invisible violence. How does it "plot and give figurative shape to formless threats whose fatal repercussions are dispersed over space and time" (Nixon 10)? Thus, below, I extend Nixon's analysis of literary texts to cinema. Whereas Nixon's primary preoccupations are with postcolonial writing (such as by Arundhati Roy, V.S. Naipaul, Jamaica Kincaid, and Ken Saro-Wiwo), here I am interested in how the "aesthetic strategy" (32) of a postcolonial filmmaker might mediate socio-environmental injustices that are not spectacular but instead accumulate over long periods of time. Specifically, I suggest that *IGL*'s blend of animation and live action illuminates what Banu Subramanian has called "archaic modernity"—the pervasive way religion and science and technology meld in modern India (Subramanian)—and thus, the destructive nuances of slow violence in India (Fig. 1).

In turning to cinema, I also suggest that *IGL*'s "aesthetic strategies" further eco-film scholarship's recent interests in animation. Despite the growing interest in animation by ecocritics, two characteristics do mark most of the studies so far. First, the predominant focus has been on either Western or Japanese animation (see, e.g., Heise; Murphy; Murray and Huemann; Brereton; Pike; Starosielski; Whitely). Second, animation's eco-potential is usually chalked to what Heise describes as its "combination of serious engagement with a playful style" (Heise para 1). "Play," or what I will call animation's "feel-good affect," is a common theme in current eco-animation studies.

By paying attention to affect, the bodily response a film invites, and emotion, the cognitive awareness of this response (Weik von Mossner 1), in *IGL*, I argue for animation's potential outside of this feel-good affect. Specifically, I show that in illuminating slow violence, a non-playful affect can be equally worth our ecocritical consideration. I draw attention to

Fig. 1. Artist Aditi Chitre's animation generates a spectacle of violence that threatens the lives of the villagers. In her animation, she incorporates the red arches of the Special Economic Zone's gate to also recall the religious markings of the Hindu *tilak*. This image is not in the final cut of *In God's Land*, but is used in the film's promotional materials. *Source*: Courtesy of Pankaj Rishi Kumar.

IGL's hybrid of dark, discordant animation spectacle interspliced in the documentary live action to articulate my argument. Ultimately, *IGL* not only draws attention to animation's non-playful affect—its potentials and dilemmas—but I also suggest that reading such a film adds postcolonial understandings of cinema beyond the Western/Japanese centre on which eco-animation scholars have so far focused.

"Nothing Is Happening": Live Action and Slow Violence's Unspectacular Affect

The initial minutes of *IGL* herald Kumar's embedding the story of the SEZ in its bigger spatio-temporal context. The opening title credit of *IGL* is superimposed over a charcoal-toned, black-and-white landscape drawing. All light and shadows, the drawing's composition cuts the frame horizontally in half. Objects that appear to be tree trunks stripe the upper half, as their shadows reach vertically along the ground in the lower half of the frame. Behind them, in the upper half of the frame, are what appear to be distant hills and a hazy sky through which the sun penetrates.

With a crack of muffled thunder, the credit transitions to a close-up of green stalks in muddy water. To the diegetic sound of falling rain, the camera lingers before cutting to three more equally lingering close-ups. On the last close-up, the subtitle appears, "A devout farmer spreads out his paddy to dry in the sun." The following sequence of four shots, each pulling out further and infused with the sound and presence of falling rain, reveals a verdant, rice-paddy landscape. The final shot of the sequence is of a lone coconut tree foregrounded in the rich green of the paddy fields and backgrounded by the sweep of an imposing hill, recalling the black-and-white landscape drawing of the title credit shot.

The subtitles locate this landscape in myth by recounting the story that resulted in the naming of the region as Tirunelveli, the "sacred hedge paddy" of a devout farmer. However, by immediately transitioning to an aerial view of a Google Maps satellite image, *IGL* transitions us to the technological present. The subtitle "Imam Alungulam is a small village in the Tirunelveli District, Tamil Nadu" accompanies the image of a small green section in the satellite image's lower left, which is outlined in black. Then, as a red-hatched boundary appears on the map, encircling a large brown section of the map and the green area of the village, Kumar's voice-over explains, "The Nanguneri Special Economic Zone was one of the earliest SEZs to be planned. It's been ten years now…"

Kumar's voice-over further explains that as he filmed, "[t]ruth became stranger as days passed. The government, the Vanamamalai temple, and villagers of Imam Alungulam narrated what the SEZ meant to them. Yet, one question remained, why are the gods lying?" As Kumar speaks, the buzz of a helicopter intrudes onto the soundscape, growing louder until it is all that remains to accompany the flashing red boundary on the satellite shot, which then cuts away to a moving car's street view of the SEZ's barbed boundary and its entrance of imposing red, curved, double arches.

These first few minutes of the film, with their extreme close-ups progressively zooming out to the panoptic scale of a satellite shot, and Kumar's juxtaposition of mythological time with the technologically present, clearly herald the film's preoccupation with embedding the contemporary moment into broader sweeps of space and time. At the heart of the SEZ land acquisition is an old tension between the upper-caste Brahmanic Vanamamalai temple and the lower-caste farmers and goat herders of the

Imam Alungulam village. Though denied by the temple swami in his on-screen interview, this tension is amply clear in many testimonies by the local villagers, who are angry and worried that the land from which they earn their livelihood has been claimed by the temple and sold as part of the SEZ.

In multiple on-screen interviews, elderly villagers recount temple authorities' insistence that the villagers are tenant farmers who must pay rent for the land or describe how villagers are restricted from entering the Vanamamalai temple and humiliated by Brahmins, who demand obsequiousness. Villagers' anxiety is revealed in one dramatic live-action moment of a local festival—a temple representative, body painted with turmeric, ash, and vermilion to signify his divinely possessed state, dances with his entourage of drum beaters and devotees through the village streets. Meant to be out blessing the village folk, he is instead accosted by a village woman who berates him and vehemently yells, "How can you show your face here? How can you be so merciless? Aren't you God?"

With these moments, Kumar highlights the Hindu caste system as a key component of Tirunelveli's landscape. While the scholarship on alliances between politics and religion in India is rich (see, e.g., Bhatt; Hansen; Jaffrelot; Van der Veer), scholars such as Madhav Gadgil and Ramachandran Guha, as well as Emma Mawdsley, have also pointed to the intersections between politics, religion, and the environment. In the case of Tirunelveli, historian David Ludden's *Peasant History of South India*, a work that Kumar accessed while doing research for his film, offers the most comprehensive look at how religion is deeply embedded in the socio-environmental lives of the region's inhabitants. Tracing Tirunelveli's history back to medieval, tenth-century Hindu nations, Ludden writes that "four types of social network shape the peasant's world: kinship, religion, state, and market interactions" (9). While each can be considered discrete, the "four networks are woven together so tightly that change in one necessarily involves the others"; for example, "worship [religion] involves loyalty [kinship] and authority [state]"[2] (10). Ludden's thesis helps trace the roots of contemporary land rights issues. The political tensions between the region's stratified communities are intrinsically eco-social entanglements, where ecological rights and privileges are woven into the social fabric of the caste system.

Following Ludden's insights, Kumar highlights four key constituents in Tirunelveli's contemporary landscape: the temple representatives, local political campaigners, the SEZ entrepreneurs, and the villagers. Each group deliberately corresponds to a caste in the Hindu hierarchy:

Brahmins (religious caste), who are the temple representatives; Kshatriyas (ruler/political caste), who are the local politicians; Vaishyas (merchant caste), who are the SEZ entrepreneurs; and Shudras (servant/peasant caste), who are the villagers. Their contemporary linkages illustrate Banu Subramaniam's concept of India's "archaic modernity," in which traditional Hindu power structures continue to influence a modern scientific, technological, and capitalist state.

Specifically, *IGL* presents the temple swami insisting that economic development is important in the local community despite contradictory testimonies from lower-caste villagers, who will be displaced. It captures live footage from local political campaigns where politicians are bedecked with garlands of flowers, which to an Indian audience makes easily apparent the allusion to gods (and religious hierarchy). Similarly, to many in Indian audiences, the sense of corruption in such politics is also evident in the scenes Kumar provides of politicians wooing voters with television sets, free eggs, and promises of gold. Not only do politicians don godly vestiges,[3] but they also use economic goods as favours to maintain power.

While such moments suggest the complicity of temple authorities and politicians, who use economic logic to justify maintaining long-held positions of power, *IGL* also captures the voices of SEZ entrepreneurs. During a fortuitous invitation Kumar received to accompany a group of visiting entrepreneurs to an on-site meeting in the SEZ, he captures the businessmen chatting and being friendly with each other. Honestly discussing the challenges of setting up business here, they share their various concerns, including those of labour costs, where they dismiss the local farmers as unsuitable. "You cannot employ them. We will have to depend on outsiders," a businessman explains, suggesting that the villagers receive "Rs. 200 per day" from a new government scheme where "they don't have to do anything. Just sit for three hours, sign, and go back."

Pitted against the villagers' own voiced concerns of hardship and dispossession, such words seem callous. Yet Kumar admits that, like his entrepreneur interviewees, many in his Indian audiences, steeped in age-old legacies of India's hierarchical social structures and couched in contemporary capitalist agendas, do not see the film's implied critique. As Subramanian writes, India's modernity covers "the old and familiar terrain" of "hierarchies of caste and class, and religion" (83), generating its characteristic brand of systemic privilege and right for the already haves.

Though *IGL* exposes the binaries of privilege and want in the contemporary moment, Kumar was dissatisfied at being unable to represent slow violence in the rough cut: "The entire history of this place, which is for me the backbone of this project, in the rough cut was represented by people talking on the screen—this happened and that happened—which was dull and boring" (personal communication).

Indeed, most of Kumar's live-action footage consists of interviews with various stakeholders. There are no instances of cinematic violence, such as beatings, or shootings, or large-scale protests. The interviewees can only describe such violence. Often presented during static "talking head" shots, their words are embedded in images of the unremarkable everyday, which may not prompt viewers to *feel* strongly, to be *affected*. As much recent work in film philosophy reminds us, this lack of affect may limit the film's ability to effect change. For example, recognising that eco-documentary films often wish to forward "consequential" messages of systemic awareness, Alexa Weik von Mossner draws on research in cognitive psychology to point out that affect and emotion are "often quicker, easier and more efficient" (46, quoting Paul Slovic xxxi) ways to get viewers to "change their rational thinking about the ecological issues they present and encourage them to become active" (45).

Kumar does not see his documentary as an explicit eco-activist film, yet as a filmmaker, he is aware that cinema's power lies in its affective and emotional resonance. To heighten such resonances, he is strategic in how he frames some of the live-action interview shots. He positions a villager by a railroad track with a train rumbling by as the man describes his impending displacement and migration as labour to a large city. He follows goat herders on grazing routes across the SEZ's recently fenced boundaries as well as on newly developed, traffic-filled roads, and silhouettes them against a darkening sky so that their forms appear faceless. Kumar often lingers on the interviewees even when they fall silent as if to let their words sink in.

Such visual moves attempt to build emotional sympathy for the villagers. Their voiced concerns are contextualised in the hardships of daily life. But, to further emotively articulate the strength of archaic modernity in Tirunelveli's present, Kumar worked with animator and friend, Aditi Chitre, to compress into cinematic time the long histories of oppression and caste/class conflict. As *The Sunday Guardian* critic Tanushree Bhasin writes, "It is in these animation sequences that Kumar's stand on the issue comes through most clearly, in the way he satirizes temple authorities or in his critique of exploitation sanctioned by religion."

"Scratches in the Land": Animating and Sounding Slow Violence as Spectacle

Chitre's animation sequences, like that of the title credit shot, are deliberately stark, charcoal-toned with deep shadows crisscrossing the landscape, and with the figures of the villagers simplistically drawn. In many shots, the villagers are dwarfed by their surroundings. In contrast, the temple and the SEZ are framed to tower over their surroundings. These symbols of power are often imbued with a harsh slash of colour. For example, the red curved arches associated with the SEZ's boundary and entrance are often re-conceptualised in these animated representations as a *tilak*, the marking of vermilion smeared on the forehead of Hindu deities and the devout. The conjoined symbolism of religion and economic progress is an *explicit* marker of archaic modernity.

The first and longest animated sequence, which begins approximately ten minutes into the film, is signalled by a live-action shot of a train edited to travel backwards on a railroad track out of the frame, leaving only the empty track and an accompanying electric pole in the image. As the animation begins, the shot switches to black and white. The subsequent animation tips the electric pole out of the ground and literally erases it from the scene (as one might with a pencil eraser). On the empty landscape, tree trunks appear, dwarfing small ant-like figures in the distance. A medium shot shows that the figures are human beings, drawn with ant-like bodies and travelling with a few similarly, simplistically drawn goats. Subtitles alert us that "a few centuries ago, this area was under Nizam's rule" when "about six families came to live here." We are clearly in the past.

This past continues to be visually presented to show the families' settlement of the land and their title to a pond, which is bestowed by the Nizam, a feudal ruler. In articulating the villagers' settlement of the land, *IGL* draws attention to a series of close-ups of hands dragging across the frame, fingers scoring black gashes across the earth. Both Kumar and Chitre note that the hands represent the villagers' statements in their life-action interviews. Kumar explains, "time and again when you talk to the villagers, their constant argument is that 'my forefathers scratched this dry piece of land and made it wet land. How dare you take it away from me?'" (personal communication). Chitre further notes the sense of toil she wanted to convey, referring to her choice of drawing the villagers as ant-like:

> The farmer folk were drawn in the shape and scale of ants because of the extreme hard work that both do, and also because it is so easy to carelessly destroy what an ant has created with its toil, without even knowing what repercussions it has to face—much like the farmers facing the giant nexus of the temple, politicians, and corporations. (Personal communication)

This "giant nexus of temple, politicians, and corporations" figures in the animation sequences as a monstrous behemoth morphing out of the hills in the landscape. Part rock, part machine (with satellite disk–shaped head, marked with the red curved arches), part giant cockroach–legged beast with a cavernous furnace for its belly, this nexus' thick, muscular tongue gluttonously eats up evidence of the villagers' ownership of the land (a copper plate provided by the Nizam) as well as the land's resources—its rice fields and coconut trees. The villagers feed the monster sacks of harvested grain, only to find themselves denied access to it. At the end of the sequence, with temple chants on the soundtrack, *IGL* provides us with a shot from within the creature's belly; the villagers peer in at us from outside, even as crosshatch bars descend to keep them out. The next shot is from outside; in long shot, we see the monster, the villagers barricaded outside its barred belly, and the landscape barren and smoking. This shot dissolves to a live-action one of the Vanamamalai temple's colourful façade, with the chants continuing on the soundtrack. The resemblance to the animated monster is unmistakable. The subsequent live-action shots of the temple, villagers loading sacks of grain and heading towards its entrance, as well as the voice-over of an interviewee recalling that "India got independence in 1947. But our village was bonded to the temple," reiterate the longer, older legacy of unequal power relationships introduced in the animated sequence (Fig. 2).

In effect, via animation, the film attempts to *spectacularly* show the wreckage caused by Tirunelveli's juggernaut of archaic modernity. While Kumar is insistent that his film is not an activist film, the condensing of centuries of unequal power relations into a few cinematic minutes, where villagers are visually overwhelmed by the "devilish" (Kumar's term) monster, serves to "plot and give figurative shape to formless threats" of slow violence (Nixon 10).

As cognitive film philosophers remind us, plotting such tensions is a familiar cinematic trope to engage audience allegiance for specific characters (Smith; Smith). In addition, the lack of voice-of-God or interviewee narrators during the animation sequences demands viewers to actively

Fig. 2. Still from *In God's Land*'s animated sequence, revealing artist Aditi Chitre's "giant nexus of temple, politicians, and corporations" morphing out of Tirunelveli's hills. *Source*: Courtesy of Pankaj Rishi Kumar.

process what they see, hear, and *feel* into cognitive sense. Kumar explains, "From the beginning we were clear, the animation had to be emotionally driven" (personal communication). Thus, while *IGL*'s live-action sequences introduce the tension of socio-environmental injustices, the animation's careful choreography of stark, black-and-white design with gashes of colour and scaled contrasts between villagers and the "nexus of temple, corporate, and political" power heighten the sensory effect of imbalance. We are, in eco-film scholar Adrian Ivakhiv's words, "struck" by "the thick immediacy of cinematic spectacle, the shimmering texture of image and sound" (ix).

While the dynamic, scaled, and stark imagery is spectacular, its effect is enhanced and complicated by *IGL*'s discordant non-diegetic sonicscape. For example, as we watch hands scratching the land, the repetitive, sawing sound that accompanies these close-ups generates a scraping dissonance. Though the visual narrative prompts us to root for the villagers working to "quench" their land, the sound of their toil is unpleasant; its discordance seems to even signify violence to the land. The volume and presence of such "unnatural" mechanical sounds get louder and more insistent in sequences that feature the monstrous nexus, pushing us to emotionally associate the temple, the SEZ, and its related turbocapitalism with sensations that are affectively unpleasant. Kumar confirms, "We wanted the track to be very problematic; a track that can disturb the serene landscape" (personal communication). In choosing a "jarring" sonic score, Kumar and Chitre deliberately heighten a sense of violence that is hard to characterise if the film depended only on its live-action diegetic imagery and sound (Fig. 3).

Through these affectively intense animated sequences, one might argue that *IGL* fits well with eco-film scholars Robin Murray and Joseph Heumann's assertion that "the genre of animated film gains power because it challenges expectations of art, film, and narrative" (2). Specifically, as the epigraph suggests, the animations "offer a contrast to the rules of live-action" (Murray and Huemann, 92), rules which, in the case of realist documentary live action, would suppress the film's ability to adequately capture a sense of violence.

Aware of the "wild possibilities that animation as a medium opens up" (Chitre, personal communication), *IGL* uses its presence to enhance what viewers are familiar with in their everyday, "real" worlds. Kumar points to how Chitre "saw hours and hours of my video footage. She said, 'I want to understand the texture of this place; the landscape of this place, how people look'" (personal communication).

Fig. 3. Still from *In God's Land*'s animated sequence, close-ups of villagers' hands scratching the land. *Source*: Courtesy of Pankaj Rishi Kumar.

Such immersion in live-action environments is a common practice amongst animators as it gives them a *feel* for the settings that inspire the affective potential of their own creations (Whitley, 2014; Brereton). For example, Graham Walters, producer of the Disney animation classic *Finding Nemo*, explains, "what you have to do in animation—any kind of filmmaking, really—is direct the audience's eyes to where you want them to go, to not just soak in the environment, but *feel* it" (qtd. in Cotta Vaz 23, and re-qtd. in Whitley, 2014: 155). Thomas Newman, the musical director of *Finding Nemo* and other animations like Pixar's *WALL-E*, similarly explains the importance of choreographing feeling when scoring animation—it's "a lot of what animation is about. It's feeling followed by transition, followed by more feeling, and typically in, like, little five-second bursts" (Siegel).

Through such audio-visual layering, one might argue, as eco-film critic Pat Brereton does when describing Pixar's *Up* (2009), that such mainstream Hollywood animations generate "a rollercoaster ride of emotions" (191). However, as Brereton further suggests, through promise of positive endings and "a reflexive array of musical registers that speak to 'childhood innocence'" (193), such popular films generally forward an overall upbeat vibe—a *feel-good* affect, with the potential to inspire positive ecological sensibilities. Like Brereton, David Whitley (2014, 2008) highlights the ecological promise of Disneyesque animations, commenting on how the brightly coloured and vibrant saturation of their hyper-real settings often enhance our sense and appreciation of the natural world. Though Ursula Heise looks at Japanese anime and French stop-motion animation, instead of Disney or Pixar, she too argues for the positive, feel-good affect of eco-animation. Drawing attention to filmmaker Sergei Eisenstein's notion of "plasmaticness"—animation's film-world of flexible beings and environments that can mutate against the rules of reality—she suggests that the genre "playfully explores ecological adaptation, resilience..." (305). One of the few ecocritics to look at animation as not necessarily playful is Sean Cubitt, who considers Canadian animator Chris Landreth's 2004 *Ryan*, a deeply dark portrait of the Canadian animator Ryan Larkin, who at the time the film was made was living as a hobo on the streets of Montreal. Cubitt is interested in *Ryan*'s self-referentiality, which, in turn, he argues, is important to draw our attention to the material resources that construct cinema (104).

While the savvy renditions of hyper-real environments we see in Hollywood's animations often wow viewers with their "natural look" and technological slickness, *IGL*'s dark charcoal illustrations appear to subvert any such lionising of cinema or nature. Instead, like with the literal, pockmarked, gaping holes in the animated *Ryan*'s screen presence that Cubitt discusses, in the scratches, shadows, and unevenness of Chitre's drawings, we are presented with a sense of imperfection and systemic socio-environmental violence. Thus, as a point of ecological entry, while the mainstream Disney animations and Japanese anime discussed by most ecocritics often use the cinematic–technological marvel of animation to present a nature that is often threatened (and threatening) yet also ultimately regenerative and regenerating, *IGL* resists such reassurances.

In its foreboding representations of human–nature relationships, it forecloses any feel good messages that might lull viewers into ecological complacency. Such a take is not only rare in mainstream Western/Japanese animation explored by most ecocritics, but also somewhat unique in India's expanding animation landscape, which is heavily influenced by Disney's conventions. While such influences reach back to the industry's roots (and the government-funded propaganda shorts of the 1950s), American studio outsourcing to a cheap Indian labour force in the 1990s fueled the country's domestic animation market and strengthened Western animation conventions (Lent; Pandyan). The lion's share of India's contemporary animation targets the country's vast children's audience and, much like Disney and Pixar, encodes an overall feel-good affect. Such an affect becomes especially pertinent to our discussion when we also acknowledge that much of mainstream Indian animation highlights India's Hindu myths (Lent; Pandyan).

Many see the mainstream meld of Disneyesque aesthetics and Hindu mythology as an expression of glocalisation—the local indigenisation of global forces—and thus, a force of positive cultural sovereignty. For example, Lent writes: "Indian parents like these 'mytho-cartoons' because they introduce ancient tales to a generation they feel is losing touch with its 5000 year heritage" (110). However, given such mytho-cartoons' strong cultural themes of Hindu tradition (including those of patriarchy and implied caste and class designations) and their heavily commercialised presence (on television and through brand marketing),[4] one can also fault them as complicit expressions of archaic modernity and its insidious slow violence. Despite the paucity of existing scholarly attention about Indian animation, such a critique is not unreasonable, particularly when viewed in the context of a number of media studies that implicate other types of mainstream Indian television and film as expressions of contemporary religious nationalism (see, e.g., Rajagopal; Kaur Dhillon and Gwynne; Guneratne).

By foregoing any sense of feel-good affect in favour of the jarring imagery and sound that portray the progressive razing of Tirunelveli's land and its ant-like inhabitants, *IGL* adopts a distinct aesthetic strategy that defiantly denies ecological complacency and resists complicity with the neoliberal project of archaic modernity.

"Disturbing the Serene Landscape" and the Perceptual Dilemmas of Slow Violence

So what does *IGL* ultimately leave viewers with?

If, as Weik von Mossner writes, emotion and affect are "central to the perception of risk, rational thinking, and decision-making in our everyday lives" (46), then viewers can take the "emotional force" of *IGL*'s animations to see and *feel* a "submerged story of injustice" (Nixon 280). Both sonically and visually intense, the animations add a level of narrative plot and character tension that generate a cinematic spectacle out of the seeming quiescence of the current SEZ controversy. They "disturb the serene landscape" that marks the live-action sequences of everyday, ordinary experiences, such as villagers working in their lush fields, herding goats across barbed fences erected to designate the SEZ lands, or discussing politics under the leafy spread of banyan trees. While the live action also captures other slightly more dramatic moments, such as the temple swami, festivals, political campaigns, and survey trips by businessmen to the SEZ, in order to heighten the affective resonance of these live-action moments, *IGL* animations portray Tirunelveli's slow violence as spectacularly accumulated and dangerous. These sequences visually bleed into and out of the "real world" (e.g., the live-action shot of the railroad tracks transitioning to its animated replica, or the animated monster of archaic modernity morphing into the entrance of the Vanamamalai temple). The discordant sonicscape also serves as an uninterrupted aural, affective bridge between the two modes of representation, helping viewers recognise the temporal relevance of violent pasts and futures (as animations also show the SEZ developed) to the present.

To further demonstrate how time is continuous with deep impacts, Kumar also experiments with his live footage. Specifically, a repeated motif throughout the live-action sequences is the presentation of shots in rewind mode. Here Kumar's intentions are twofold. First, there is a direct critique of traditional caste practices, where the untouchable caste are required to walk backwards when in the presence of upper-caste Hindus. Second and related, Kumar's critique is of India's current development trajectory: "I wanted a sense right then and there—we keep talking about progress, but what does it actually mean?" (personal communication).

By pairing this rewind motif with the animations, the answer seems clear—progress is dangerous and disruptive; it is regressive. However, by coupling these rewind live-action shots with another repeated live-action editing motif—that of spinning camera shots—*IGL* provides a more nuanced response to the question. The answer isn't easy to grasp. Slow violence is dizzyingly complex in its long temporalities. We can't tackle it by simply being "anti-SEZ":

> It's quite possible that the SEZ will be thrown out in a couple of years and on that piece of land, they'll then come up with another concept, but the basic tenet of that concept will also be displacement. How does it matter to them [sic] if it is a SEZ or a formula 1 track? What was more important to me was the displacement part of it, the power equation. (Kumar, personal communication)

Though suspicious of progress, *IGL* also argues against simplistic environmental messaging by refusing to sentimentalise the past (even in the villagers' relations with their land). It thus escapes Emma Mawdsley's label of "neo-traditional environmentalism," which projects "a historically inaccurate past environmental harmony and an equally inaccurate past social harmony, both of which have been contaminated and degraded by a foreign cultural domination" (384). By acknowledging that recent Western capitalist forces are not the only culprits of Tirunelveli's inequities, *IGL* resists standardising Indian environmentalism as a nationalistic project.[5] Instead, through its insistence on giving the farmers and goat herders of Imam Alungulam the ability to speak alongside other stakeholders, *IGL* often presents environmental protest through polyvocal and often contradictory expressions. For example, Kumar captures a group of villagers arguing over who they will vote for in the upcoming elections, with obvious differences being expressed. Such presentation forces an "unevenness in the delivery of such activism," complicating what Jyotirmaya Tripathy (building on Gadgil and Guha's work) suggests are "uncritical and unproblematic theorizations of such activism that often reduces environmentalism to questions of life and livelihood" (72). As the animations suggest, the villagers' ability to maintain control over their lands is central to their life and livelihoods; however, as the live-action testimonies and sequences help us recognise, this is not simply a story of homogenously helpless victims. In the complex arena of India's development, individuals react differently—from the woman vehemently berating the temple representative, to villagers enlisting lawyers in their case, to those more resigned to migrate to the cities. Recognising these heterogeneous responses and understanding the multi-layered dimensions that inform Tirunelveli's contemporary moment are powerful first steps for viewers if they wish to tackle the realities of India's complex environmental issues. However, even as *IGL* forwards this message of complexity, there is a caveat in its cinematic delivery.

IGL's aesthetics—from its charcoal-toned animations, to its disharmonious soniscape, its rewind and spinning shots, its attention to Tirunelveli's complex religious hierarchy, and its overlapping eco-social intermeshes—are affectively "jarring." While one might argue that such intended cinematic dissonance is part of Kumar's intent—after all, the situation is frustratingly layered—it can confuse viewers (as it did some of my own students). Not only is the immediate visual and aural affect discombobulating, but to make sense of the multiple parts, viewers have to work especially hard. As scholars such as Dan Flory write, when viewers are confronted with a cinematic world with which they are uncomfortable, viscerally and cognitively, it is easy for them to "pop-out" of the experience and resist its messages (again, as it did with some of my own students).[6] Perhaps equally important to the dynamics of "popping out" of *IGL* is the film's unrelenting seriousness, its overall effect of doom-and-gloom, which, as eco-film scholars such as Nicole Seymour and David Ingram suggest, often discourage viewers from wanting to watch ecocinema. Describing most ecocinema as "serious business," (61) Seymour suggests that many view it as "too much like taking medicine" (63), and thus shy away from it.

Overall, *IGL* is not an easy film to watch. There are no easy feel-good moments to fall back on. Yet, despite the perceptual dilemmas presented by Kumar's aesthetic choices, even my students who "popped out" of the experience found themselves drawn back in by the film's spectacular animations, which "moved" them "most immediately and directly" (Ivakhiv xi). Ultimately, when we dig deep to analyse its aesthetic strategies, *IGL*'s representation of slow violence is well worth our ecocritical attention. In its postcolonial dimensions, *IGL* helps us look beyond the

Western/Japanese centre to extend our current eco-readings of affect and emotion in animation. As such, it makes a call for more eco-attention to both animation that is not necessarily "feel good" and to animation that gives voice to those outside the mainstream, global cinema market. It also takes up the challenge of scholars, such as Nixon, Mawdsley, Gadgil and Guha, and Tripathy, to complicate our understanding of Indian environmentalism as simplistic stories of subaltern victims confronting Western-imported capitalist violence. In India's contemporary politics of archaic modernity, we recognise instead long histories of socio-environmental injustices meted out through the caste system and actively resisted in heterogeneous and complex ways.

Notes

1. The SEZ's designation is specifically for non-extractive industry, something like a food processing centre, or a pharmaceutical company, yet the closest viable economic hubs (the sea port of Tuticorin and the city of Madurai) are each approximately 100 km away, while the state capital, Chennai, is a good 600 km away.
2. I have provided the bracketed terms to help clarify Ludden's links.
3. Indian democratic politics, while well-known for its alliances with more fundamental religious groups, has an additional, peculiar religious element. Specifically, politicians such as Jayalalitha, a contender in the elections at the time of Kumar's shooting, are well-known movie actors who have played roles as Hindu deities in India's popular film industry. Jayalalitha is famous because of her career start in Tollywood, the Tamilian version of Hollywood.
4. For example, as seen in Ram Mohan's popular *Ramayana: The Legend of Prince Rama* (1992), or the popular animated feature *Hanuman* (2005).
5. See Mawdsley's "Hindu Nationalism, Neo-traditionalism, and Environmental discourses in India" for an overview of how some environmentalists, such as Vandana Shiva, have been faulted for such problematic, neo-traditional narratives.
6. Flory is specifically concerned with how the film medium affects a viewer's moral perceptions and generates Humean or "imaginative resistance," which in turn fuels a "popping out." In this case, *In God's Land* does not present a Humean "alternative morality" from the one its environmentally attuned viewers might have, but instead complicates its one main morality lesson, that of destructive development, with many competing perspectives. Here, I am less interested in Flory's focus on imaginative resistance, even as I draw on the importance of his discussions regarding a viewer's resistance to a film based on its affect.

Acknowledgements I would like to thank the students of my spring 2013 Environmental Film class who helped launch me on this analysis. I also sincerely extend my thanks to Pankaj Rishi Kumar and Aditi Chitre, who were both immensely generous with their time and insights. Finally, I am also grateful to Rayson K. Alex for his patience as an editor and, as always, thankful to Matt Beehr for his sharp copy-editing.

References

Anwar, Mohammad Amir. "New Modes of Industrial Manufacturing: India's Experience with Special Economic Zones." *Bulletin of Geography. Socio-economic Series* 24.24 (2014): 7. Print.

Bhasin, Tanushree. "God's Land Bought and Sold for 'Progress.'" *The Sunday Guardian*. 19 Jan. 2013. Web.

Bhatt, Chetan. *Hindu Nationalism: Origins, Ideologies, and Modern Myths*. Oxford: Berg, 2001 Print.

Brereton, Pat. "Animated Ecocinema and Affect: A Case Study of Pixar's Up." *Moving Environments: Affect, Emotion, Ecology and Film*. Ed. Alexa Weik von Mossner. Waterloo, ON: Wilfrid Laurier UP, 2014. 181–200. Print.

Chitre, Aditi. Animator. Personal Communication (email), 19 Sep. 2014.

Cubitt, Sean. "Ecocritique and the Materialities of Animation." *Pervasive Animation*. Ed. Susanne Buchan. New York, NY: Routledge, 2012. 94–114. AFI series.

Flory, Dan. "Race and Imaginative Resistance in James Cameron's Avatar." *Projections*. 7.2 (2013): 41–63. Print.

Gadgil, Madhav, and Ramachandran Guha. *This Fissured Land: An Ecological History of India*. 2nd ed. Oxford UP, 2012 Print.

Guneratne, Anthony R. "Mediating the Rise of Neo-Nationalism in India: Television, Cinema and Carnival." *Social Identities* 4.2 (1998): 263. Print.

Hansen, Thomas Blom. *The Saffron Wave: Democracy and Hindu Nationalism in Modern India*. Princeton, NJ: Princeton UP, 1999 Print.

Heise, Ursula. "Plasmatic Nature: Environmentalism and Animated Film." *Public Culture*. 26.2 (2014): 301–17.

In God's Land / KADAVULIN NILATHIL. Dir. Pankaj Rishi Kumar. Production PSBT, 2012. Video recording.

Ingram, David. "The Aesthetic and Ethics of Eco-Film Criticism." *Ecocinema Theory and Practice*. Ed. Stephen Rust, Salma Monani, and Sean Cubitt. New York: Routledge, 2013. 43–62. Print. AFI Series.

Ivakhiv, Adrian. *Ecologies of the Moving Image: Cinema, Affect, Nature*. Waterloo, ON: Wilfrid Laurier UP, 2013. Print.

Jaffrelot, Christopher. *Religion, Caste, and Politics*. Delhi: Primus Books, 2010. Print.

Kannan, K. P. "Interrogating Inclusive Growth: Some Reflections on Exclusionary Growth and Prospects for Inclusive Development in India." *Indian Journal of Labour Economics* 50.1 (2007): 17. Print.

Kaur, Dhillon Narinderjit, and Joel Gwynne. "Saffronizing Bollywood Cinema." *Film International* 12.2 (2014): 47–57. Print.

Kumar, Pankaj Rishi. Filmmaker. Personal Communication (phone interview). 7 Aug. 2014.

Lent, John A. "Animation in South Asia." *Studies in South Asian Film & Media* 1.1 (2009): 101–17. Print.

Levien, Michael. "Regimes of Dispossession: From Steel Towns to Special Economic Zones." *Development & Change* 44.2 (2013): 381–407. Print.

Ludden, David E. *Peasant History in South India.* Princeton, NJ: Princeton UP, 1986. Print.

Mawdsley, Emma. "Hindu Nationalism, Neo-Traditionalism and Environmental Discourses in India." *Geoforum* 37 (2006): 380–90. Print.

Monani, Salma, and Andrew Hageman. "Ecological Connections and Contradictions: Penguins, Robots, and Humans in Hollywood's Nature Films." *Hollywood's Exploited: Corporate Movies, Public Pedagogy, and Cultural Politics.* Ed. Tony Kashani, Rich Van Heertum, Ben Frymer, and Anthony J. Nocella II. New York: Palgrave McMillan, 2010. 203–18. Print.

Murphy, Patrick. "'The Whole Wild World was Scrubbed Clean': The Androcentric Animation of Denatured Disney." *From Mouse to Mermaid: The Politics of Film, Gender, and Culture.* Ed. Elizabeth Bell, Lynda Haas, and Laura Sells. Bloomington, IN: Indiana UP, 1995. 125–36. Print.

Murray, Robin, and Joseph Heumann. *That's all Folks: Ecocritical Readings of American Animated Features.* Lincoln, NE: U of Nebraska, 2011 Print.

Nixon, Rob. *Slow Violence and the Environmentalism of the Poor.* Cambridge, MA: Harvard UP, 2011. Print.

Pandyan, Kanakasabapathy. "The Coming of Age of Indian Animation." *Journal of Asian Pacific Communication* (John Benjamins Publishing Co.) 23.1 (2013): 66–85. Print.

Pike, Deidre. *Enviro-Toons: Green Themes in Animated Cinema and Television.* Jefferson, NC: MacFarland Press, 2012.

Rajagopal, Arvind. *Politics After Television: Religious Nationalism and the Reshaping of the Indian Public.* Cambridge, UK; New York: Cambridge UP, 2001. Print.

Seeds of Dissent. Dir. Pankaj Rishi Kumar. Production PSBT, 2009. Video recording.

Seigel, Robert. "Scoring the Screen: Thomas Newman and the Music of Shawshank Redemption" *National Public Radio* Interview. 22 Dec. 2014. Web.

Seymour, Nicole. "Irony and Contemporary Ecocinema: Theorizing a New Affective Paradigm." Ed. Alexa Weik von Mossner. Waterloo, ON: Wilfrid Laurier UP, 2014. 61–80. Print.

Smith, Greg M. *Film Structure and the Emotion System*. Cambridge, UK; New York: Cambridge UP, 2003. Print.

Smith, Murray. *Engaging Characters: Fiction, Emotion, and the Cinema*. Oxford: Clarendon Press, 1995. Print.

Starosielski, Nicole. "'Movements that are Drawn': A History of Environmental Animation from Lorax to Fern Gully to Avatar." *International Communication Gazette* 73.1–2 (2011): 145–63.

Subramaniam, Banu. "Archaic Modernities: Science, Secularism, and Religion in Modern India." *Social Text* 3 (2000): 67. Print.

Tripathy, Jyotirmaya. "Indian Environmentalism and its Fragments." *Ecoambiguity, Community, and Development: Towards a Politicized Ecocriticism*. Ed. Scott Slovic, Swarnalatha Rangarajan, and Vidya Sarveswaran. Lanham, MD: Lexington Books, 2014. 71–83. Print.

Van der Veer, Peter. "Religion, Secularism and National Development in India and China." *Third World Quarterly* 33.4 (2012): 719–32. Print.

Weik von Mossner. Alexa. "Emotions of Consequence? Viewing Eco-Documentaries from a Cognitive Perspective." *Moving Environments: Affect, Emotion, Ecology and Film*. Ed. Alexa Weik von Mossner. Waterloo, ON: Wilfrid Laurier UP, 2014. 41–60. Print.

Whitley, David. "Animation, Realism, and the Genre of Nature." *Moving Environments: Affect, Emotion, Ecology and Film*. Ed. Alexa Weik von Mossner. Waterloo, ON: Wilfrid Laurier UP, 2014. 143–158. Print.

Whitley, David. *The Idea of Nature in Disney Animation*. Aldershot, UK: Ashgate, 2008. Print.

Transnational Perspectives on Land Ethics: *Elemental* and *Not My Land*

Pietari Kääpä

A group of international tourists embarks from a bus in what appears to be rural Estonia. They visit a communal farm inhabited by a group of Russian nationals who have been tending to the land for more than 30 years. The tourists gaze at the group in curiosity as a large passenger plane flies overhead, preparing to land at the nearby international airport. The juxtaposition of a wide variety of different registers—from ecological harmony, implied by the commune, to resource-intensive means of air travel, from local authenticity to touristic awe at such a means of existence—guides viewers to consider a range of perspectives on societal relationships with nature.

Meanwhile, a different group of individuals alights from a bus in rural Canada to protest against the construction of the Keystone XL Pipeline at the tar sands, one of the largest deposits of crude oil on the planet. The construction of the pipeline threatens both the local environment and the indigenous people that inhabit the land. In addition, it also poses a serious threat to global climate stability via the carbon emissions that would enter the atmosphere from the use of its crude oil in the production and consumption chain. In different ways, the global and the local intertwine in both instances, not least in the way that the issue of local environmental

P. Kääpä (✉)
Media Management, University of Stirling, FK9 4LA Stirling, Scotland

R.K. Alex, S.S. Deborah (eds.), *Ecodocumentaries*,
DOI 10.1057/978-1-137-56224-1_3

protection emerges as a distinctly global one, in terms of its moral and political relevance.

These examples come from two very different films. The first is from the documentary film *Not My Land* (2011), an Estonian production directed by Aljona Surzhikova. The film, a humanitarian tale focused on individuals relocated to Estonia to work at the Dvigatel factory during the Soviet occupation, is also "a green documentary," to use the vocabulary employed by Helen Hughes (2014) to describe documentaries with an environmentalist bent. On the one hand, the film's concerned depiction of the demolition these plots face from airport expansion is a clear case of conservationist logic, seeking to preserve nature from humans. But on the other hand, this is a film focusing primarily on human lives, directed by an individual with a distinctly complex "Soviet" history, with parents from the Ukraine and Uzbekistan. Due to this dual role, *Not My Land* facilitates a complex meditation of ideological transmutations from a communist system of central control and "egalitarian" distribution of land to a more capitalist system governed by the dual logic of private property and individualism, even as it also tries to provide an environmentally powerful exposition of contemporary Estonia.

The second description comes from a very different type of film, both in its geographical context and in its mode of green rhetoric, even though it shares a similar concern for nature's role in contemporary society. *Elemental* (2013), directed by Emmanuel Vaughan-Lee and Gayatri Roshan, chronicles the struggles of three individuals with ecological challenges in three countries on different continents—India, Canada, and Australia. Whereas *Not My Land* is more concerned with a nostalgic sense of attachment and lament for the changing lifestyle of a group of people, *Elemental* aims to encourage activist responses to the problems facing humanity "living in the end times," to quote the title of a recent book by Slavoj Žižek (2011), as well as highlighting the similarities such problems have the world over. Despite their differences, both films provide an exploration of the ways in which humanity uses the land for its own power politics, as well as the often detrimental consequences such myopic actions have. They are especially significant in illuminating the complex ethical relationships humanity maintains with its immediate ecosystem, as well as with the planetary environment as a whole.

To explore the implications of such local and global depictions of environmental relations, this chapter takes Aldo Leopold's elaboration of the land ethic in his seminal *Sand County Almanac* (1949) as its key starting point. As an environmentalist take on philosophical ethics, the notion of the land ethic is useful for characterising the sort of complex evaluation of humanity's fraught relationship with the biotic environment, as it emerges in these films. In the words of Leopold, the concept of the land ethic "enlarges the boundaries of the community to include soils, waters, plants, and animals, or collectively: the land." By expanding the scope of the community beyond its anthropocentric imagination, the concept suggests a fundamentally ecological take on interconnectivity, not only between humans but also their ecosystemic presence—one that is characterised by a reciprocal, but often extremely unbalanced, flow.

The concept of the land ethic has frequently been appropriated for film studies. This is not surprising as the more anthropocentric variations of the land ethic are key to many film narratives. Particularly, the use of an economic or utilitarian land ethic—that is, *those* seeking various ways to position and use the land for the benefit of humankind—features in narratives of conquest and ownership of land and has been a key part of such essential cinematic tropes as the Western (Carmichael 2005) and the road film (Brereton 2012). As *these* emphasise the conquest of "undiscovered" lands and work symbolically to construct individualistic self-realisation or narratives of nation building, their environmentalist contributions must remain debatable. Other writers focus on more egalitarian forms of the land ethic, which often take the form of anti-colonialist critiques through emphasis on, for example, minority politics. Such an egalitarian approach emphasises the significance of the equal distribution of resources, as well as the use of conservationist strategies to preserve the land; however, it also works in critical terms as a way to interrogate hegemonic appropriations of indigenous land. In cinema, this appropriation often takes place symbolically as either narrative material or audiovisual signifiers—approaches which have been critiqued widely, from Berilla's ecofeminist readings of Robert Redford's films (2010), to Kiu-Wai Chu's (2013) exploration of CCTV's (Chinese Central Television) use of the nation's river systems in its documentary productions.

Underlying all these perspectives is the understanding that the way in which nature is used in film is conducive to a plethora of power games, from identity politics to governance of nations, where the land supports the specific sociopolitical arguments the films conduct. In short, the majority of cinematic depictions of the land are utilitarian versions of the land ethic due to the limitations of their world views. They often struggle in attaining the ecological dimensions evoked by Leopold's land ethic, especially in terms of placing the human as part of the much wider biotic community. This is not entirely surprising as capturing such a perspective through the means available to cinematic production is difficult due to its inevitable constitution as an inherently anthropocentric activity, a fact that has been addressed by many writers, including Adrian Ivakhiv (2012) and Scott McDonald (2004).

In exploring these issues, I focus on the diverse modes of the land ethic in *Elemental* and *Not My Land* to address the ways cinematic depictions aspire to work environmentally, but often fail to do so because of their inherent limitations. As they describe land politics as a question of both local and global relevance, I will investigate their arguments in a transnational framework, especially in terms of how they appropriate land for national narratives. This is significant as nations are one of the primary ways in which humanity seeks to realise its political goals, which, in this case, it does by appropriating land for sociopolitical gain. Most considerations of identity and culture, for one, rely on both visually striking depictions of nature and iconographic framings of landscapes utilised to testify to the uniqueness of a particular national culture. In the book *Transnational Ecocinema* (2013), Tommy Gustafsson and I develop this suggestion to argue that a transnational perspective can provide much for ecocritical analysis of films. First of all, the significance of this approach emphasises the need to adopt a critical perspective on the self-evidence of nations, especially when they are conceptualised as natural for a wide range of politically nefarious reasons. Second, it necessitates considering the inevitable cross-border flow of environmental issues as well as the international and domestic regulatory mechanisms established to confront them. Simply put, as befits an ecological land ethic, we have no inherent right to ownership of the land, and thus the focus on a transnational understanding of environmental considerations encapsulates the refusal to settle on anthropocentric simplicities drawing on nature for narratives of nationhood. In adopting a transnational perspective on the land ethic, *Not My Land* and *Elemental* are both explored as cases illustrating humanity's complex ecological role, as well as the conventional but persistent narratives of selfhood in much of non-environmental cinema that seek to avoid addressing this complexity, and which thus avoid acknowledging human complicity in environmental destruction.

Not My Land: A Political Land Ethic?

Of the two examples discussed in this chapter, *Not My Land* is perhaps the most immediately anthropocentric case, especially as its most immediate concern is to construct a distinctly humanitarian argument. Its focus on the farm allotments, which used to be "the people's" property during the Soviet occupation of Estonia, provides it with an easily identifiable political project. Accordingly, land was considered property that is "provided" for the people by a totalitarian regime, an act that ultimately aims to consolidate a deeper sense of identification with this regime. As Estonia was able to gain its independence in the early 1990s, the rights to the land became contested. The rationale for the commune gaining access to this land was, in fact, a matter of historical injustice as the oppressive Soviet occupation wrestled the land away from the Estonians in the name of a vehemently flawed and destructive misappropriation of an international communist ideology. Thus, any of the settlers' rights to this land—at least outside of ecocritical questions about any anthropogenic rights to land ownership—are debatable. However, before addressing the implications of these problematic political connotations further, it is necessary to investigate a key concern the film raises about these settlements—their role as a means of egalitarian sustenance for individuals, regardless of their national identity.

Not My Land is clearly on the side of the group of settlers, who have had to devote considerable time to cultivating the land. We hear about their relocation orders from their statements as well as from protest songs, which make it clear that the problems here are more than just about affectionate attachment. The settlers are forced to live in a state of liminality as they are not able to, nor do they want to, go back to Russia. Yet they are unable to get Estonian passports as the language tests demanded for citizenship are difficult even for natives. The residents state that in contemporary Estonia, it is impossible to survive on the meagre state pension provisions. Their ability to survive depends on the food they gather from the land and from the herds of animals, which make these settlements close approximations of self-sustaining farms. Sustainability is a key idea here, enforced by frequent shots that pan to the sky, with planes thundering over the land. It is clear that the film considers the farms as a more sustainable way of life than the loud and polluting consequences of the airport. In one particularly symbolic shot, the film contrasts a beetle on the surface of a dry highway with the interviews of the farmers, suggesting a parallel between two ways of life teetering on the edge of destruction. In providing such a stark contrast between sustainable communities and capitalist initiatives, the arguments of the film are reminiscent of many of the basic tenets of Joel Kovel and Michael Lowe's "Ecosocialist Manifesto":

> Ecosocialism insists upon redefining both the path and the goal of socialist production in an ecological framework. It does so specifically in respect to the "limits on growth" essential for the sustainability of society. These are embraced not, however, in the sense of imposing scarcity, hardship, and repression. The goal, rather, is a transformation of needs, and a profound shift toward the qualitative dimension and away from the quantitative. From the standpoint of commodity production, this translates into a valorization of use-values over exchange-values—a project of far-reaching significance grounded in immediate economic activity. (http://www.iefd.org/manifestos/ecosocialist_manifesto.php)

The film's emphasis on creating contrasts between the green tenements and the barb-wire walls that separate them from the airport, as well as the airplanes that rumble overhead, makes it an intriguing reflection of the manifesto's emphasis on locally produced goods and the avoidance of the hyper-consumption of imported products from faceless, multinational conglomerates. One character even explicitly vocalises what amounts to an ecosocialist argument. He suggests that humans have to find ways of living that are based on sharing resources and abiding by a global sense of communality. According to him, this is a problem to do with national organisations which exploit ordinary citizens in favour of capitalist gain, though what is left unsaid is the collusive behaviour between multinationals and governmental regulators. This argument for a universal socialism challenging nations is very reminiscent of a Marxist call to arms and seems to fit in with the film's political inclinations. The repeated use of protest songs on the soundtrack, with overhead helicopter shots of the plantations, also contributes to this sense of explicit socialist rhetoric and its take on the land ethic—in short, it provides the film with an argument that ties in with a utopian form of ecosocialism.

In creating this argument, the film asks crucial questions concerning human appropriation of land, showing us the sociopolitical movements that seize the land as part of the project of consolidating nationhood, as well as the temporary rule of such appropriations in the "longue duree." Yet, much of *Not My Land* is based on affectivity with the human subjects it depicts, which is more in line with facilitating anthropocentric connections rather than unravelling them. What seems to be largely missing from the film is any comprehension of the rights of the land itself. While such calls for respecting nature may be somewhat utopian and reminiscent of some of the more extreme ecological perspectives, it is a necessary question to ask in relation to such explicitly environmentalist films as *Not My Land*. Furthermore, such questions can also lead us to ask intriguing questions relating to what exactly is being said through these environmentalist claims. This is especially relevant as the film emphasises historical fluctuations and political arguments that challenge particular political claims to soil—that is, what it sees as the corruptive role of the contemporary Estonian state—and proceeds to present these claims as part of an anthropocentric world view of fundamental myopia.

The film certainly consolidates a comprehensive ecosocialist argument through visual and aural means, but as Joel Kovel, one of the key figures of the ecosocialist "movement," reminds us, such arguments are not an end in themselves as they often contain substantial practical and ideological contradictions. For Kovel, this is more of a question to do with unequal distribution of resources as well as of overpopulation, both of which do not feature in *Not My Land*. While the film emphasises concerns integral to ecosocialist rhetoric, including both the tragedy of the commons and post-statist approaches to social order, their role as both environmental and political signifiers needs to be looked at more closely. These concerns could be taken as critiques of the lingering colonialist presence or memory of the Soviet Union in Estonia, but such arguments are never clearly posed. Moreover, as the film seems to almost celebrate the "glory days" of the occupation, with its nostalgic archival film of the lives of the commune members when they were young and employed in the state-controlled factories, the use of the land ethic as part of the film's ideological constructions becomes muddled. Through this, it seems to fall victim to a rather simplistic utopian conception of USSR socialism.

As much of the emotional power of the film is targeted at the contemporary capitalist order, what is problematically absent here is a balanced critique of the system that brought the individuals to Estonia in the first place, destroying any notion of balanced economic development in the Baltic states. The focus on displaced Russian nationals, the nostalgic gaze at the communist era, the wholesome characterisation of the commune and its more "authentic" use of the land, and, finally, the lack of care for communities and the environment by the modern, capitalist Estonian state seemingly positions the Soviet strategies as more ecologically sound than those of the "modern" West. All that seems to matter to contemporary Estonia is profit and expansion, a reading complemented by the inclusion of the aforementioned lengthy tourist scene in a film of 52 minutes.

While criticism of capitalism's detrimental effect on the environment is certainly commendable from an ecological point of view, the lack of balance in the film is problematic in its uses of the land ethic and its seeming aspirations to an ecosocialist approach to political rhetoric. While the film does aspire to the kind of humanism present in the social ecology of Murray Bookchin, suggesting that challenging relations of domination concerning human over human also extends to human over the natural environment, *Not My Land* does not seem to take this to its most efficient conclusion. Instead, utopianism and a limited understanding of the political context pervade the film. Social ecologists have suggested that the most effective form of ecocriticism can emanate from addressing systematic injustice and critiquing contemporary land politics. The film provides a searing exposition of capitalism, but as a humanist text with an undertone of political bias, it shows many of the problems involved in environmental, documentary film production with a focus on human subjects. By choosing an environmental theme to support an ideopolitical argument, the film seems to reduce the environment to its normative role as a tool for human politics. By establishing this relationship, any sort of progressive rhetoric it seems to emit becomes part of an anthropocentric logic that only validates established power paradigms.

Nowhere is this more apparent than in the conclusion as we hear one of the residents musing about the impermanence of contemporary life and how humanity lives in imbalance with nature. Overhead, we see darkening clouds, which provide a stark counterpoint to the demolition of the settlements taking place beneath them. The harmony, both of a socialist and an ecological type, is shattered, and as the dark skies overtake the idyll, only planes, both as vessels of technological might and as symbols of the capitalist order, pass overhead. It seems that technology has surpassed the organic in human thinking and we are now entering uncertain times. This final combination of aural and visual calamity, combining ecological doom and social upheaval, constructs a powerful argument criticising contemporary ways of life. But even as it does this, *Not My Land* harkens back to a nostalgia that ignores the realities and history of its time. Ecosocialism, in such a limited form, does not take into account the historical past of the contemporary condition and remains an ideologically problematic take on rethinking the human–land relationship.

ELEMENTAL: ECOCOSMOPOLITAN TAKES ON GLOBAL ENVIRONMENTALISM

Returning to the question of the transnational, let us begin from the premise that *Not My Land* illustrates the tensions that arise when land is appropriated for political projects. While nations, for sure, are an enigma rather than a taken-for-granted element of the film, this enigma is never explored to such a degree as to shed light on the environmental implications of such appropriations. In practice, the film never once questions its own use of the land ethic in constructing its narrative, nor does it address the potential implications this has for its political content. This is not surprising as similar tactics are used by a variety of "ecodocumentaries" with a transnational scope. These range from films that seek to confront the environmental costs arising from the cultural habits of other nations (e.g., *The Cove*, Psihoyos 2009) to ones that involve questions of global environmental relevance (e.g., Davis Guggenheim's *An Inconvenient Truth*, 2006). The former's narrative follows a group of mostly American ecoactivists chasing Japanese dolphin fishermen (for further discussion on this film, see Selheim 2013) and constructs a transnational framework premised on universal moral values to question Japan's national customs. In *An Inconvenient Truth*, US Senator Al Gore travels the world to demonstrate the very real need to curb carbon emissions, even as the film instigates a more problematic discourse concerning a white, Western male lecturing the rest of the world on how to run their lives. Similar examples are plentiful, from Mika Koskinen's documentary critiquing the Finnish foresting company Stora Enso's exploitative policies in China in *The Red Forest Hotel* (see Kääpä 2014) to documentaries about photographers capturing the desolate nature of industrial China in *Manufactured Landscapes* (Bachwai 2007).

In contrast to these explorations of Western environmentalists "bettering" the global ecoscape, *Elemental* takes a more self-consciously transnational perspective on the universality of environmental issues. It focuses on three individuals who struggle with ecological challenges in their specific cultural contexts. By focusing on three countries on different continents—India, Canada, and Australia—the film shows us not only the necessity of engaging ecopolitics on a local scale but also the fact that similar problems face activists the world over. The focus of *Elemental* is on three individuals: Eriel Deranger, an environmental activist from the indigenous Dené population in northern Canada who raises the issue of minority rights;

Rajinder Singh, an Indian politician and campaigner for conserving the Ganges River; and Jay Taylor, an inventor from Australia working on new types of energy production technology emulating patterns from nature. As with *Not My Land*, the focus here is on the individuals in charge of the activist campaigns. This is an entirely understandable position considering the history of the singular hero in ecodocumentaries and the fact that conveying stories of activism can create identifiable role models and facilitate the efficient conveyance of a film's message to audiences.

By having three distinct cases focusing on protagonists from different cultures and social standings, *Elemental* provides a more complex take on transnational ecoactivism. This is especially evident when compared with *Not My Land* and its combination of national politics and universal ideological networks. Here, *Elemental* is emphatically planetary in its argumentative scope, which focuses on the ways that environmental problems may often start from very specific local contexts, but soon flow to ecosystems outside of the initial context. These are entirely reciprocal, however, as environmental problems do not recognise the borders of nations, at least in terms of toxins and other harmful substances entering water and air systems external to those from which they originated. The three individuals, whose stories the film narrates, are all activists for causes with distinct transnational dimensions. They all campaign against environmentally hazardous practices with the aim of changing the ways in which public and legislative authorities view nature, as well as its resources. Furthermore, these individuals are emphatically aware of the ecosystemic nature of the issues they campaign about as each of them frequently positions the problems they face in a more profoundly global context. Deranger, of the three the most "stereotypical" activist, campaigns against the harvesting of crude oil from the tar sands in northern Canada, but she also makes it clear that the use of these deposits will have considerable global environmental consequences. Thus, she collaborates with the non-governmental organisation (NGO) Rainforest Action Network (RAN), which seeks to protect forests from the exploitative practices of a range of multinational corporations. She is also an extremely outspoken protector of the rights of the indigenous people of the area and often falls foul of authorities and sponsors. This leads to the termination of her collaboration with RAN and she then must continue trying to find ways to fight the destructive impact of the pipeline, using independent methods.

Echoing Deranger's aims to challenge the planetary harm caused by business interests, Rajinder Singh is conducting his own battle against the large-scale pollution of the Ganges River. He frequently asserts that the state of the river is largely due to "democratic" governmental policy controlled by capitalist interests, and we follow his mission to educate communities about cleaning the river as well as ensuring policy changes for its survival. One of the key problems is that "the capitalist democracy" in control of India is planning to construct dams to increase energy production. The dams alter the river system and contribute to pollution, thus challenging the survival of communities along the river. By showing the need to act simultaneously with local communities and the continental ecosystem, *Elemental* emphasises the importance of extending any ecosocialist rhetoric beyond the immediate community, not only in the abstract rhetorical terms found in *Not My Land* but in practical application as well. With several of the local communities, many of whom are distinctly dubious of his motivations, having encountered state corruption before, Singh argues for both local and national management of the river to ensure that it continues a sustainable existence.

The final part of the complex global connectivity of *Elemental* takes place in Australia, where the innovative green technology developed by Taylor draws on nature's cyclical patterns to combat increasing climate change. Much of the film focuses on a prototype he intends to sell for use in the Chinese market, involving a jet engine-like power source to create air funnels that can disperse pollution in the atmosphere. The technology is designed to be energy efficient, making it a competitive commodity for markets like China, where such innovative products meet with considerable interest from a government subsidising the renewable industries. Much of this part of the film focuses on his negotiations with investors, who can see the potential in the new technology but also express considerable concern over its functionality. While it may seem somewhat disingenuous to highlight technology in an environmental film such as this, the way these arguments are posed provide an alternative vantage point on ecosocialism.

The film's take on human ingenuity illustrates the complex standpoints that ecosocialism takes in balancing the complicated reciprocal relationship between humanity and nature. The argument starts out from the premise that humanity has to consume a wide variety of resources in order to sustain itself. Ecosocialist arguments seek to conceptualise perspectives whereby these consumption practices could be limited in relation to how much and how the process of consumption takes place. Its view is not to do away with consumption in some utopian scenario but to start to envisage ways of making this depletion more sustainable. Green technology can thus be conceptualised as a means to a more ecosocialist life perspective and Taylor's role in *Elemental* seems to largely corroborate this.

A wider question on ethics still pertains as developing new products to mitigate the damage caused by centuries of overconsumption only addresses one small aspect of the bigger picture. On a more substantial level of environmental activity, perhaps it would be more important to address the process of consumption in its own right. Instead of a focus on developing new products to sustain the status quo, would it not make more sense to address the situation in its own right? *Elemental* incorporates this argument by repeatedly showing us scenes illustrating Taylor's love of nature when he escapes from city life to find a solid connection with the earth in the "wilderness." These scenes can be interpreted in a variety of ways. On the one hand, they clearly present an alternative to his work, which seems to mostly focus on patching up the damage already inflicted on the natural world. But in providing such idyllic ruralism as an answer to the problems of contemporary society, the film falls back on a clichéd opposition between authentic rural life and distanced city existence. Such binary polarisations lack the finesse of an ecosocialist perspective on addressing the ways contemporary life works today. From this perspective, an elopement to pure nature is only a fantastical diversion into abstract utopias, with little real value in addressing such issues as global warming or climate change. The development of green technology, on the other hand, is a practical incentive with a realistic strategy to counter humanity's impact on the environment.

Elemental's perspective on utopian green politics is also present in *Not My Land*, which, while not explicitly stating so, argues for collective farming as a reasonable alternative to current regular production and consumption processes. From an ecocritical viewpoint, both *Elemental* and *Not My Land* make sense and contribute valid perspectives on alternative means of living. The problem is that these are consistently presented as alternative perspectives rather than as ways of challenging mainstream issues. The politics of resource depletion are clearly a part of these narratives but neither of the two films successfully addresses their inevitable assumption of a range of anthropocentric values. The point is not that the films ought to try and construct some sort of authentic environmental perspective but that they fall back on conventional clichés in environmental rhetoric. This seems more obvious in the case of *Not My Land* as it does not position itself as an ecofilm with a distinct environmental message. However, this argument is a lot more difficult to make in the case of *Elemental* as it clearly positions itself as a film with an environmentalist agenda. The problem in this case comes from *Elemental* not addressing the more fundamental systemic questions over human consumption processes. To begin from a more ecoethical viewpoint would necessitate putting this question at the centre of any discussion on sustainable views of living. In fact, *Not My Land* could be considered more successful from this perspective in that it provides a valid alternative to resourcing sustenance. But its main problem arises in its framing of its ecological arguments as part of a political project. *Elemental* comes from the opposite viewpoint, in many ways, as it provides a case for global human solidarity in the guise of environmentalism. Yet, the human is at the centre of the film's perspective, leading us to the inevitable question of whether ecocinema ought to be striving for ways to reconsider this relationship. But how can we displace the human from the centre? Is it even possible to do this without alienating much of the core audience, who are, of course, inevitably those of the human persuasion?

Without getting too carried away by these philosophical conundrums, it is fair to say that cinema is a human activity aimed at other humans. Thus, it is an inevitably anthropocentric machine, to quote the succinct definition expressed by Adrian Ivakhiv (2010), that addresses a very specific allocation of the biological spectrum. Its arguments are inevitably coloured by this perspective and this means that any radical solution it offers has to be done within these limitations. But this does not mean that it is not able to strive to exceed its inevitable anthropocentricity. Of course, many films try to minimise the presence of the human in more experimental works in cinematic environmentalism, but they fail to largely address the bigger picture of the rationale for the existence of these works, which is that other humans see them and incorporate their messages into their own ways of thinking.

Thus, if we go back to considering films such as *Elemental* and *Not My Land* through an ecosocialist perspective, we can see their real areas of impact. It simply is not possible for these films to exclude the human element. Their work has to be aimed at a level of behavioural change that can affect their target audience. We need to start seeing the rationale of ecological technology as a means to finding an alternative to the fossil fuel industry. These ideas come through powerfully in both films. To requote an earlier definition of ecosocialism as concerning "a valorisation of use-values over exchange-values," both films depict situations where ecosocialist behaviour takes tangible form. Neither try to do away with the human in some sort of impossible quest for a non-human means of representation. Instead, they argue the case for a productive technological way to try to curtail humanity's negative impact on the planet. But even as they do this, they fall victim to a very human fallacy that testifies to the impossibility of pure ecocinema.

While many of the critics of ecosocialism suggest that some of its principles are utopian in orientation and thus not able to work in practice, we need to consider the alternative. They propose a more pragmatic take on the economic potential of green technology, especially as some of this technology can be costly and its implementation requires substantial investment and support from private enterprises and governments. This is an argument that situates economic logic at the forefront and does not consider other alternative ideological positions. Much of the work done by environmentalist organisations has had to adopt the rhetoric of economics in order to make their work possible. For example, the film industry in the UK today has to report on its environmental footprint to gain funding from Arts Council of England. The logic of economics is, therefore, an inevitable part of any contemporary sustainability effort. By incorporating this discourse into the film, *Elemental* takes into account the pragmatic tensions that underlie greening efforts. Yet, far from being a simplistic acceptance of green capitalism, the film emphasises Taylor's green credentials and focuses on his need to escape to nature in the Australian outback. While, cynically, these scenes could be interpreted as a form of greenwashing intended to balance the more explicitly business-minded sections of *Elemental*, the ways in which the film captures this sort of emphatic green thinking is complemented by placing this section amidst the other narratives of the film. This creates a compendium argument that sees environmentalism as a politically complex planetary phenomenon.

To further explore these perspectives on ecosocialism, we now turn to the ecocosmopolitan discussion of Ursula Heise (2007). For her, ecocosmopolitanism comprises an approach that acknowledges both the local and global dimensions of ecological issues. Simply put, environmental problems inevitably flow over national borders, even as they are often initiated and managed by local or national communities. This creates a push–pull dialectic that understands environmentalism as a reciprocal phenomenon of a range of factors influencing the ecosystem of the planet.

Ecocosmopolitanism also encapsulates many of the arguments of *Elemental*. Deranger, Singh, and Taylor are all motivated by their concern for the land and its appropriation by commercial interests. The conflicts in all these cases are clearly both local, in that they concern local environments and communities, and ecological, in that they do not remain bound by any territorial borders. The mining of oil from tar sands will harm not only the ecosystem of the local community but also impact on North American ecology and beyond. Simultaneously, the flow of industrial pollution and human waste in the Indian water system causes not only misery for local inhabitants but also within the wider context of the continental water basin. Through this, they embody the basic premise of ecocosmopolitanism as they envision "the imbrication of local places, ecologies, and cultural practices in global networks that reconfigure them according to a logic that recent theories of globalization label 'deterritorialization'" (Heise 2007: 210), indicating the ways environmental issues can transcend place-bound considerations.

Such instances in *Elemental* also tie in with ecocosmopolitanism's reciprocal tendencies by acknowledging that the most effective, as well as restrictive, strategies are often locally implemented. Heise refers to them as processes of reterritorialisation, which revert the focus to the specific contexts in which ecosystemic transformations are felt. In *Elemental*, an example of an effective strategy involves activists like Singh talking to local communities to make sure they use the river systems in sustainable ways. In contrast, an example of the restrictions imposed by such strategies emerges in the legal frameworks of nations obfuscating Deranger's ecoactivism. As seen from these contrasting takes on ecosystemic activity, a cinematic form of ecocosmopolitanism is about illustrating the reciprocal ways environmental problems operate both locally and globally and thus necessitate a form of politics that transcends local considerations to demand a response that works on a much wider level. Ecocosmopolitanism, thus, characterises a fundamental part of *Elemental*'s rhetorical pose and captures well the ways in which contemporary environmental rhetoric must be conducted.

Not My Land also makes use of aspects of ecocosmopolitan logic as the community's existence is premised on transnational considerations. As with the push–pull depictions of *Elemental*, *Not My Land* also highlights the positives and negatives of global ecosystemic connectivity. From the perspective of the film, the airport connecting Estonia to a global society is not seen as a positive facet of contemporary social order. The threat posed by the expansion of the airport is a case of deterritorialisation, here viewed as leading to disenfranchisement from the land. In contrast, the film sees the most value in local roots, thus creating a case of reterritorialisation. Here, localism is the antidote to the unbalanced and inconsiderate means that both displace individuals and disadvantage the environment, all conducted in problematic rhetorical terms that pose very specific forms of political longing as the roots of its ecosocialist point of view.

Considering *Elemental* from an ecosocialist perspective is not difficult either. It clearly makes a case for a planetary sense of responsibility, especially when all three of its stories combine. Such evocations are not unique either as many a globe-hopping documentary—from *An Inconvenient Truth* to *The 11th Hour*—construct similar modes of discussion that not only touch on key geopolitical locations, but also use these touchstones to construct a planetary ecosystemic argument. Yet, the positive connotations of such perspectives should not be taken for granted. Elsewhere, I have questioned the implications of ecocosmopolitanism (2013). This is to do with the term cosmopolitanism and its inevitably positive connotations as a term emphasising transcendence and levity. Questions here revolve around whether such an approach can illustrate the complex dynamics of ecosystemic relationality, especially when the emphasis seems to be on "the big picture" rather than on the tangible consequences and gains at ground level. Certainly, the picture seen in *Not My Land* is much more focused in its implications than *Elemental*, yet the scope of the latter allows it more potency as well as avoidance of politically contentious problems. The question is not really a matter of deciding which film is "better" in its use of the land ethic; indeed, these texts have different aims in mind and cannot be compared in a literal way. While both *Not My Land* and *Elemental* emphasise the threat of modern progress for "authentic" communities (especially in the case of the former), the understanding of *Not My Land* remains bound to its context, whereas the picture that emerges in *Elemental* is much more optimistic. This is especially true in the case of inventors like Taylor, who use the resources available from emerging, progressive economies to combat such destruction. What these cases highlight is the dexterity of the ecocosmopolitan approach. By illustrating the land ethic through this approach, it becomes clear that ascribing value judgements on a given text's take on the land ethic is a process of scale, of seeing the land ethic as both context and argument specific.

Conclusion

In order to understand the applications of the land ethic in *Elemental* and *Not My Land*, this chapter has used two ideological frameworks—ecosocialism and ecocosmopolitanism—to assess the complex ways in which humanity represents its nature dependence, and mobilises this discourse for specific sociopolitical goals. By deconstructing the politicised project of nationhood, this chapter has used the notion of the transnational as a means to critically interrogate such imperatives. By doing this, I found that an ecocosmopolitan approach, despite its problematic take on ecosystemic connectivity, has a more consolidated sense of environmental responsibility than sticking to restrictive nostalgic imaginaries. By evoking a shared environmental responsibility—a global ecosocialist ethos—both films ultimately testify to the difficulty of balancing human stories with environmental rhetoric. Even as they evoke environmentalist arguments, they also conceptualise nature as property. This, in turn, gives rise to an anthropocentric way of thinking of (human) being in the world. Because of this inevitable anthropocentricity, the way in which the land ethic emerges in specific texts is best considered as a means to highlight the complex, shifting modes of identification with land, as well as humanity's place within it. This is the case even with clearly environmentalist texts such as *Elemental*. Thus, they are open to interpretations over both their environmental and their ideological messages—a notion that is especially important for films that include alternative political content as part of their environmental ideas. Ultimately, this discussion of the diverse modes of the land ethic as a cinematic trope illuminates the concept's significance for environmental film studies, as well as its considerable and necessary malleability.

References

An Inconvenient Truth. Dir. Davis Guggenheim. Participant Media, 2005. Film.

Berilla, Beth. "Engaging the Land/Positioning the Spectator." *Framing the World: Explorations in Ecocriticism and Film*. Ed. Paula Willoquet-Maricondi. Charlottesville: U of Virginia, 2010. 116–34. Print.

Brereton, Pat. "Appreciating the Views: Filming Nature in Into the Wild, Grizzly Man, and Into the West." *Ecocinema Theory and Practice*. Ed. Salma Monani, Stephen Rust, and Sean Cubitt. London: Routledge, 2012. 212–21. Print.

Carmichael, Deborah. *The Landscape of Hollywood Westerns: Ecocriticism in the American Film Genre*. Salt Lake City: U of Utah, 2006. Print.

Chu, Kiu-Wai. "From *My Fancy High Heels* to *Useless* Clothing: 'Interconnectedness' and Ecocritical Issues in Transnational Documentaries." *Transnational Ecocinema: Film Culture in an Era of Ecological Transformation*. Ed. T. Gustafsson and P. Kääpä. Bristol: Intellect, 2013. 127–44. Print.

Heise, Ursula. *Sense of Place and Sense of Planet*. Oxford: Oxford UP, 2007. Print.

Hughes, Helen. *Green Documentaries*. Bristol: Intellect, 2014. Print.

Ivakhiv, Adrian. "An Ecophilosophy of the Moving Image: Cinema as Anthrobiogeomorphic Machine." *Ecocinema Theory and Practice*. Ed. Salma Monani, Stephen Rust, and Sean Cubitt. London: Routledge, 2015. 87–105. Print.

Kääpä, Pietari. "Transnational Ecocinema: Charting an Expansive Field." *Transnational Ecocinemas: Film Culture in an Era of Ecological Transformation*. Ed. Tommy Gustafsson and Pietari Kääpä. Bristol: Intellect, 2012. 35–59. Print.

Kääpä, Pietari. *Ecology and Contemporary Nordic Cinema*. New York: Bloomsbury, 2014. Print.

Kovel, Joel, and Michael Lowe. "An Ecosocialist Manifesto." 25 Aug. 2014. Web. <http://www.iefd.org/manifestos/ecosocialist_manifesto.php.>

Leopold, Aldo. *The Sand County Almanac*. New York: Ballantine Books, 1970. Print.

Manufactured Landscapes. Dir. Jennifer Baichwal. Foundry Films, 2007. Video recording.

McDonald, Scott. "Toward an Eco-cinema." *Interdisciplinary Studies in Literature and the Environment* 11.2 (2004): 107–31.

Punaisen Metsän Hotelli. Dir. Mika Koskinen. Luxian Productions, 2011. Film.

Selheim, Megan. "Towards a Political Economy of Ecodocumentary." *Interactions: Studies in Communication and Culture* 4:2 (2013). 129–47.

The Cove. Dir. Louis Psihoyos. Diamond Docs and Skyfish Films, 2009. Film.

The 11th Hour. Dir. Davis Guggenheim. Appian Way, Green Hour, 2010. Film.

Žižek, Slavoj. *Living in the End Times*. London: Verso Books, 2011.

Communal Indian Farming and Food Ecology: A Reading of *Timbaktu*

Pat Brereton

Overview

The importance of provocative and insightful documentaries has never been more pertinent, especially with increased global concern over food poverty and obesity, resulting from highly saturated food products and increased factory farming. Most especially, there is a prevailing need to influence perceptions and behaviour in supporting alternative modes of living, while facing up to the global challenges of climate change.

Scholars are beginning to address this of late with a number of edited volumes, including Willoquet-Maricondi (2010), Rust et al. (2013), and Weik von Mossner (2014), alongside scholarly journals such as *Environmental Communication*. These are all creating a growing body of research, exploring a range of concerns and testing out effective strategies and theoretical models to assist in this challenging area of study.

This chapter will focus on a short Indian documentary titled *Timbaktu* and will explore its use of direct address and nature visualisation. It foregrounds a persuasive cautionary tale concerning the overuse of pesticides and the general abuses of factory farming methods, while endorsing organic communal modes of farming production. The overall message of the documentary speaks to agriculture as a global industry, recounting its

P. Brereton (✉)
School of Communications, Dublin City University, Glasnevin, Dublin 9, Ireland

R.K. Alex, S.S. Deborah (eds.), *Ecodocumentaries*,
DOI 10.1057/978-1-137-56224-1_4

drive towards increasingly monocultural cash crops, rather than embracing a more balanced and organically driven method of land husbandry. The overall trajectory of the text highlights the paradigm shift required in facing up to climate change, through the practical application of more sustainable modes of food production.

Film audiences, I would ideally like to imagine, must become critical spectators of the growing range of films and documentaries. In turn, the mass media will hopefully propose some answers to environmental questions, especially calling attention to the most appropriate and best models of farming and food production. Religious film scholars, for instance, talk of the need for spirituality in film, to which I would add the growing necessity for a form of environmental enlightenment and sacredness, helping audiences to "see differently." This notion of seeing differently can certainly be applied to a deep ecological re-imagining of the visual and aural techniques of film and is, I will argue, evident in this poetic documentary. Furthermore, this media analysis approach emanates from the proposition that a tremendous amount of moral thinking, feeling, and perception is undertaken when watching films and documentaries. It has been inferred from numerous studies over the years that for most people the mass media is the primary way in which they acquire memories and ethical attitudes towards "right living" within contemporary culture.

Food Security, Sustainability, and De-growth

All the while, it remains difficult to communicate complex environmental and ethical messages around food security, much less recalling the pervasive dangers of climate change. This is especially true as it appears politically unacceptable to radically suggest that the global economy ought to shrink, in the struggle to become more environmentally sustainable and by using less carbon-based energy sources in the longer term. Most people, especially in the West, remain addicted to unlimited wealth, affluence, and expecting freely available cheap food and energy, while embracing a more destructive form of conspicuous consumption that certainly does not stack up against the challenges towards creating environmental sustainability (see Jamieson 2012).

As a concept, sustainability came to prominence back in the 1980s and has gained broad acceptance far beyond environmentalist circles. There remains, however, considerable disagreement about how to achieve a sustainable way of life, much less agree over what this might incorporate.

Scholars frequently quote the 1987 Brundlandt Report, "Our Common Future," to anchor a comprehensive definition of the concept. The idea of communal and social sustainability, in particular, which this documentary calls for, implies a two-way dialogue across what can broadly be characterised as nature and human culture. Yet, the very notion still remains a slippery concept, which, in terms of food, can be characterised as essentially promoting a "balanced" form of agriculture that does not seek to overexploit the land for short-term gain.

The ultimate manifestation of sustainable farming as a practising mode of living would ideally embrace an organic model of communal practice, like that lovingly displayed in *Timbaktu*. Promoting and popularising such radical solutions to food production, by calling for a stoical, resilient, or frugal form of de-growth, or at least not supporting artificially induced growth through the use of dangerous chemicals and a drive towards monocultural cash crops, involves a more historically conceived and balanced production–consumption model. This remains difficult to visualise, much less promote, on film (see Retzinger et al. 2008; Dobrin and Morey 2009; Murray and Heumann 2012; Rust et al. 2013; Weik von Mossner 2014). *Timbaktu* serves as a progressive antidote around the inherent potentiality in human nature towards getting the environmental balance right, by both respecting the sacredness of the earth and striving to be in harmony with nature.

Western lifestyles, in particular, will have to change and adopt globally conceived environmental justice models, especially taking into account the huge imbalances of poverty and deprivation across the globe. It is often suggested that the so-called Third World peoples, like indigenous cultural representations of American natives (see Brereton 2005), have an intrinsically sacred respect for natural phenomena, such as land, mountains, rivers, as well as trees; all of which are particularly valorised in this documentary. Everything remains personal and alive for such a vivid, nature-based culture. This philosophy of everything being connected echoes a first principle of environmental ethic and resonates with *Timbaktu*, which also speaks to a rural way of life that has been apparently lost in the West.

As affirmed by numerous scholars (see Brereton 2016; Ivakhiv 2013), it must be constantly acknowledged that far too much eco-filmic discourse and scholarship comes from a very narrow Western-centric, philosophical framework. Indian culture and the celebration of alternative ecologies of nature, such as *tiNai* (to be discussed later), will, in the future, become a major source of ecological scholarship. Unfortunately, at present, we have a long way to go before we can harvest the fruits of such a trans-global vision, alongside calling on other spiritual and discursive ways of engaging with nature. A good start has been made, however, with studies such as *Transnational Ecocinema* (Gustafsson and Kääpä 2013) and *Culture and Media: Ecocritical Explorations* (edited by Rayston K. Alex et al. 2014).

No more are crudely constructed binaries that

> polarise postcolonial/Western; ideology/ecology as simple uni-directional frames or debates helpful in articulating, much less constructing progressive forms of engagement. Instead such binaries need to be re-appraised and re-coded, as world cinemas fluctuate across the three-dimensional nexus of local/global, indigenous and trans-national polarities and feed into more complex debates around the periphery/centre and erstwhile reductive divisions between Third, First and other world cinematic outputs. (Brereton 2016: 219)

Furthermore, by developing a more productive and reflexive sense of place through civic identity and environmental connectivity, as highlighted in this documentary, it can help us re-imagine and even promote a more benevolent and cooperative series of protocols across our polarised planet earth.

This utopian trajectory involves a radical transformation of public perceptions and behaviour around food production and consumption. Some evidence of this tentative transformation can be garnered from the growth of organic farming and the slow food movement in the West, together with case studies, such as this documentary. Following the methodologies developed and extended in *Environmental Ethics and Film* (Brereton 2016), which alternatively examines Hollywood fictional narratives, this chapter will focus on the very recent factual story of a sustainable farming commune.

Rayson K. Alex most usefully suggests that a postcolonial form of ecocinema should examine this complex interrelationship between the cultural, the natural, and the sacred, which he calls "T-documentaries" (Alex et al. 27). *Timbaktu* certainly corresponds with this trajectory in setting up a complex and symbiotic interrelationship between all stakeholders within a food production community and involving an alternative and more benevolent economic system, a constructive social structure, an educational structure, and even a holistic medical structure. These all help to create a sustainable and progressive community that will become a beacon for civil society everywhere to follow and emulate. Textual analysis of this insider's vision also serves as a sounding board for contrasting scenarios and hopefully conjuring up a range of audiovisual stimuli to tease out potentially radical environmental and ethical hypotheses around all aspects of farming food production.

Environmental eco-scholars such as myself, working from Ireland, tend to place a remarkable degree of faith in cinema's capacity for wonder and hope and the need for close engagement with the materiality of land, nature, and lived environments. At first sight, I would have to agree that cinema seems to be a poor substitute for direct and effective engagement with a range of environmental challenges facing the modern world. However, its global reach in itself ensures that media has an important role to play in promoting awareness around environmental issues. It helps to construct new modes of popular engagement through sometimes radical re-visualisation of environments, drawing from a long romantic history celebrating the therapeutic representation and evocation of nature. The pervasive exposure and popularity of nature documentary on television and in the cinema, as well as on online sites like YouTube, where I sourced this film alongside the nature-focused growth of fictional genres, from Westerns to road movies and science fictional adventures in particular (see Brereton 2005, 2016), have all helped in keeping the environment centre stage and also in highlighting the growing tensions around food security.

A Deep Environmental Register for Documentary Film

There are at least five basic assumptions implicit in such a broadly environmental ethical perspective: (1) everything is connected to everything else; (2) the whole is greater than the sum of the parts; (3) meaning is context dependent; (4) process has primacy over parts; and (5) humans and non-human nature are one[1] (Merchant 76–78). These interweaving aspects of a deeply felt environmental consciousness around nature and culture are most explicitly manifested and explored in *Timbaktu*. Nonetheless, eco-scholars like Adrian Ivakhiv (2013) worry that a strictly aesthetic, moralistic, or overly romantic approach to environmentally framed film and documentary falls short of offering critics a sufficient toolkit for identifying and analysing various inherent contradictions. In quick response, however, I would counter that putting such issues on film is sometimes enough in itself. At least, such dramatic exposure helps get the whole debate started.

Film remains a particularly good spectatorial medium to draw attention to the value of land and wilderness and the importance of its preservation, which is incisively foregrounded in this documentary. Some Western environmental scholars wonder whether nature worship is only appealing to those who can afford it and, therefore, do eco-films simply provide a "cheap" and vicarious engagement with such otherwise expensive, Third-World, touristic, raw nature experiences? *Timbaktu* certainly calls attention to such concerns with its valorising of a robust, working-class, farming community displaying an inherent balance with nature which is, at the same time, clearly defined and anchored within economic and (in) justice paradigms. Furthermore, such a documentary addresses the question around how can the physically reviving power of the wild—recalling E.O. Wilson's notion of "biophilia" and the innate love of nature—become available to those living in relatively poor conditions in India and elsewhere?

Such deep environmental and justice narratives recall so-called debates around "First Wave" environmental analysis, with scholars like Chris Tong suggesting that viewing film ecologically "is very far from the uni-directional nature of first-wave eco-criticism" (Tong 116). It is suggested that, because of their single-minded and overarching concerns for conserving some early ecological eulogies of nature, they transcend or ignore the problems of gender, class, ethnicity, and wealth in the pervasive drive towards glorifying the intrinsic power and beauty of nature (see Selheim 129–46). Early eco-film scholarship, through its powerful appeal to the raw beauty of nature, tended to universalise, or even dismiss, such critical injunctions and barometers of power and injustice. But such tensions, of course, need to be faced head on for the multi-layered appreciation and challenges of environmental ethics to mature and develop across film studies and within the social sciences generally. By any measure, *Timbaktu* takes on all these sometimes contradictory tensions as it proceeds in close-up to examine the soil of an otherwise barren wilderness.

A Cautionary Tale of Metabolic Rift, Soil, and Postcolonial Environmentalism

John Bellamy Foster, in his study "Marx's Theory of Metabolic Rift: Classic Foundations of Environmental Sociology" (1999), contends that Marx provided a powerful analysis of the main ecological crisis of his day, while raising "fundamental issues about the antagonism of town and country, the necessity of ecological sustainability" and what Marx called the "metabolic" relation between human beings and nature.[2]

Given the increasing privatisation of nature and natural resources through (contemporary) industrial-style farming and other wealth-producing modes of development, Foster asks whether capitalism can be "reshaped" to the demands of what the environmentalists call "sustainable development." Foster's analysis leads him to conclude that the first steps towards resolving ecological problems lie in "the direction of the socialization of nature and production and the creation of a more democratic, egalitarian world order" (1999: 73). A radical model of farming practice is explicitly exemplified in the documentary *Timbaktu*. Many environmental scholars, in teasing out the tensions between left–right ideological and environmental pressures, agitate over opposition to the hegemonic dominance of capitalist methods of production as being an irresolvable utopian political dream. By all accounts, constructing environmentally and socially aware individuals is an essential first step towards successful environmental transformation. Yet, some scholars remain unconvinced, asserting that while ecocentrism, along with deep ecology and postcolonial power debates over land usage, does indeed have a role to play in this process, "nonetheless given society's unwillingness to see beyond anthropocentrism, it is necessarily an incomplete role" (Gandhi 47).

From an Irish or an Indian perspective, Foster cites Marx's interest in chemical research into the degradation of soil fertility by intensive capitalist agriculture and the subsequent "rift" involved in the system's metabolism, which he linked to the growing separation of "town and country." Foster and others favour generalising the "metabolic" or "ecological rift" to understand and counter "capitalism's war on the earth" (Foster, York and Clark). This form of degradation of the soil has, of course, particular purchase for a critical reading of farming and nature film generally, but most especially, it recalls outbreaks of famine, such as occurred in Ireland, as well as across other postcolonial countries. With a conscious de-nurturing of the land and the promotion of an unhealthy monocultural form of agriculture,[3] this regressive form of land abuse continues to the present day with the overuse of pesticides on the land. This spectre of loss of land quality has particular echoes in *Timbaktu*, alongside Irish farming films such as *The Field* (1990) and *Man of Aran* (1934), which focus on farmers' struggle to re-fertilise the land while drawing seaweed across long distances.

Embracing these divergent yet interconnecting precepts, one could suggest that the essence of environmental sustainability, alongside the historical notion of "metabolic rift" and the colonial discourse of denuding soil to maximise output, coalesces around a complex reframing and engagement with the disciplines of economics, national history, postcolonialism, and environmentalism. Alternatively, connotations of going "back to nature" and promoting a form of communal sustainability also register a level of sacrifice that promises to deliver fullness of being, through solidarity and environmental sustainability.

Speaking to Audiences and the Creative Imaginary of Film

By all accounts, environmental arguments will only have force if we physically feel them. In other words, if an argument fails to generate feelings, or does not tap into an affective range of public engagement, then it will probably not persuade its intended audience. This is why the creative imaginary of fiction and poetic faction remains so important in mobilising and framing public opinion, while calling on the extensive power of emotions towards affecting audience responses. Such affective arguments, it is suggested in the literature, help motivate audiences by inducing deep feelings, including satisfaction, pleasure, excitement, interest, anger, or even distress. If they generate no feelings at all, they are unlikely to be persuasive.

Eco-film scholars seek out and promote this more productive strategy and strive to extend the deep ecological mantra that "everything is connected," which in turn hopefully promotes a proactive environmental ethical perspective. But illustrating, much less proving, that such creative imaginaries have a measurable and direct effect on audiences remains open to question.[4]

The Green Revolution and Factory Farming

The so-called "Green Revolution" in the latter half of the twentieth century facilitated a huge growth in agricultural production with, for example, new and "improved" strains of grain being developed. But at what cost to the purity and sacredness of nature and to the long-term managing of biodiversity and food security? This critical cost is clearly dramatised and highlighted throughout this Indian case study.

Having been relatively cheap for the global consumer due to mass production, food is now most certainly back on the political agenda. Using chemicals to simply improve growth and secure overall high yields is having detrimental effects on human welfare, with particular concerns across India, where the problem is exacerbated. Human food safety has to be the number one concern for agriculture and society generally—as Rachel Carson uncovered with regards to DDT usage back in the early 1960s, which in turn helped to kick-start the environmental movement.

Furthermore, the global right to food remains the preeminent food justice argument seen across all parts of the globe. Observing the failure of the market-driven global system to meet this core human right, the food justice lobby has brought together an array of arguments around food and the environment, food and health, food and labour, and food and hunger, as well as focusing on how food is grown, processed, and distributed within a justice framework. The right to food, then, is not simply about food grown sustainably, but also about food grown locally, and about good and just food.[5]

The Cultivation of Food Justice

The environmentalist mantra—act locally, think globally—is also a food justice truism. When food is produced through an industrial system and distributed through a global supply chain, the inputs into food production, processing, and shipping generate enormous environmental stresses that cause pollution of the land, air, and water, pushing health burdens back onto farm workers and other food producers. These tensions are set up and aired at various levels in *Timbaktu*. When food production, including most importantly the necessity of maintaining the precious resource of creating seed for future crops, is simply commodified, when it crosses borders and oceans as the literature affirms, its environmental and carbon footprint is greatly increased and this also affects food security, as this documentary creatively testifies.

Threats to the food supply chain represent only one of these "dreaded outcomes," but because food can speak so vividly to both short-term and long-term human needs, alongside recalling other manifestations of environmental change, the search for food amidst a world of scarcity becomes a frequent motif in post-apocalyptic, fictional narratives and in documentaries like the one discussed in this chapter.[6]

Food is imagined not only as a necessary sustenance for corporeal needs, but also, according to Jean P. Retzinger (2008) in a special issue of *Cultural Studies*, as a liminal cultural symbol of life and death, nature and culture, human and non-human. For food not "only shapes our bodies, but it structures our lives, fashioned daily rituals and helps to mark significant rites of passage. Food connects us to others—both directly, through shared meals, and culturally, through shared tastes" (Retzinger 370).

Timbaktu helps to elevate alternative ideas and agendas around food, while suggesting radical change towards embracing organic farming. Yet it must be recognised that this radical concept of food justice remains underexplored in the media, particularly when it comes to analysing the phenomenon as a potential spark for a new social movement within the organisation of agriculture. For example, Tim Lang and Michael Heasmon argue that "there is rising evidence of injustice with the food system," which they characterise as "the maldistribution of food, poor access to a good diet, inequities in the labour process and unfair returns for key suppliers along the food chain" (Gottlieb et al. 4). This sense of injustice is usefully explored and turned on its head in *Timbaktu*, with small farmers and workers in the commune becoming empowered by reversing this imbalance and creating an alternative and more equitable system of food production. A Western-dominant food system produces a predominantly high-fat, high-salt, high-sugar, supersized diet, which also targets consumers of low-income communities and communities of colour, where the lack of access to fresh, affordable, healthy food has increased prevalent health disparities—see, for instance, *Supersize Me* (2004). Added to these problems is the overuse of pesticides to sustain this level of overproduction of processed food. Consequently, factory producers have to add "minerals" back into such overly processed foods, a paradoxical dilemma which this documentary effectively brings to our attention.

Farmers as Stewards of the Land

The inherent benefits of food grown locally, as embodied in *Timbaktu*, is strongly connected with Aldo Leopold's idea of the "land ethic," which is further expanded by Wendell Berry and other proponents of the fundamental and universal ethics of farming stewardship. In this way, clear distinctions can be drawn between industrial modes of agriculture, which view the land as a commodity designed simply to facilitate high production outputs, as opposed to local or sustainable food models which regard the farmer as a land steward, fostering an environmentally sustainable agri-ecological ethos, in which land is instinctively understood to be connected to biological diversity and ecosystems. This, in turn, promotes a place-based perspective, with the farmer remaining a central part of the interconnected nexus between nature and community. By any measure, the ever-popular farmers' market concept, or the evocatively visualised agricultural commune in this documentary, meets all these criteria, namely food grown locally and sustainably, which tastes good and fulfils a food justice ethic.

From a Western perspective, however, environmental critics constantly speak of the pollution of the natural environment, which is often traced back to (big) farmers and thereby questions their ethics of land husbandry, together with their role as stewards and protectors of the rural landscape. This inherent tension has been well articulated by scholars like Wendell Berry, alongside national and international organisations designed to protect endangered rural habitats. But when such protection and symbiotic relationship with the land collapses, resulting in the abuse of natural resources to simply maximise monocultural food production, it is left to imaginative individuals and pioneers like C.K. Ganguly, who tells the story of *Timbaktu*, to show a better way of instilling a more creative and sustainable engagement with land, nature, and food.

Radical Ecological Thinking: The Need for a Paradigm Shift in Agriculture

Aldo Leopold was possibly the first to articulate this environmental "Old Testament mantra" ethic by declaring: "[A] thing is right when it tends to preserve the integrity, stability and beauty of the biotic community. It is wrong when it tends otherwise" (1949: 224–25). E.O. Wilson and his controversial theory of "biophilia" (1984) insinuates that because we have evolved from nature, we still carry a part of nature in our hearts and this is where humans feel their relationship with, and responsibilities to, the land, through the complexity of human–land relations. Leopold most certainly recognised that humans have "inherently psychological affinities to the natural world, including aesthetic appreciation, emotional attachment, spirituality, and all of these affinities have evolutionary and developmental significance" (Kellert in Knight and Riedel 2002: 133).

Most incisively, in an essay titled "Carrying Capacity and Ecological Economics," Mark Sagoff (in Crocker et al.) illustrates how early environmentalists, such as Henry David Thoreau, cited the intrinsic properties of nature rather than its economic benefits as reasons to preserve it. It was believed that economic activity had outstripped not just its resource base but also its "spiritual" purpose. This spiritual purpose has all but been lost, in the West at least. Another founding father of environmentalism, John Muir particularly condemned the "temple destroyers, devotees of ravaging commercialism" who, "instead of lifting their eyes to the God of the mountains, lift them to the Almighty dollar. This condemnation certainly was not simply a call for improved cost-benefit analysis. Nineteenth-century environmentalists saw nature as full of 'divinity' and regarded its protection, less as a prescribed prima facia economic imperative than as a moral test" (Sagoff in Crocker et al. 28).

A central tenet of ecological farming could be reduced to some form of "harmony with nature," together with the recognition of "finite resources." Everything else is therefore either peripheral to or, at best, ancillary to these all-inclusive prime directive affirmations. Nonetheless, there is often little agreement through the large rainbow of green supporters on the specific means, especially the priorities and timescales, to achieve these end results. Simply looking for hope through an artificial development of holistic systems can be read as a recipe for disaster. I would suggest, however, that *Timbaktu* does not fall into this trap, remaining anchored in its explicit practical application of a holistic system of production.

Film and media generally remain an important allegorical model for "re-thinking" farming and ecology, while re-situating a sense of place and environmental identity within a natural framework. There is much fruitful potential in developing robust strategies for re-imagining our sense of identity and connectivity with the environment, especially as *we* face up to major global difficulties around climate change. Audiences need the stimulus of provocative tales like *Timbaktu* which, in turn, help provide creative imaginaries of our inherent duties and responsibility towards protecting our habitat and our future.

Frequently, I have argued elsewhere that mainstream films, with their universally engaging narratives, at least unconsciously strive to bring people and nations on the planet together within a common purpose, while tapping into a latent hunger for connectivity within a broadly conceived environmental system of values. Such films can produce, what I describe as, "creative imaginaries" and "ecological fantasies" which feed into the global consciousness of world audiences. The use of creative imaginaries most certainly extends to evocative documentaries like *Timbaktu*. While such "feel good" ecological fictional narratives can sometimes be criticised as vacuous, or even dismissed as overtly romanticised and upholding the status quo, they still remain significant, I would contend, if only for keeping the environment at the centre of human consciousness (Rust et al. 2013; Ivakhiv 2013; Weik von Mossner 2014).

At the outset, as several environmental studies recognise, it is difficult to get across deep environmental and radical ethical messages as it remains politically unacceptable in most quarters to simply insist that the global economy ought to shrink to become environmentally sustainable. Most in the West, alongside many government stakeholders across the world, appear addicted to "affluenza" (wealth and affluence) as a more destructive form of conspicuous consumerism, while embracing the long-term expectation of continuous economic growth. Lifestyles, many environment scholars assert, will have to change to accommodate necessary sustainable development and *we* need to prepare for such a radical transformation carefully.

Some might try to counter-argue that such radical evocation of a more enlightened agriculture will simply end up returning to clichéd romantic imagery, embracing a "back to nature" ethos and philosophy. In my own research, I have explored a number of studies of Irish films and documentaries which discuss farming and effective resource management using similar strategies. These include the classic *Man of Aran* of the 1930s, which evokes a primitive form of coexistence with land and sea, to more contemporary postcolonial examples such as *The Field*, where land and farming was more preoccupied with ownership and power over the land. Such films serve as useful cautionary tales, highlighting what is wrong with ownership fixations with land and the drive towards cash crops as the only mode of success. Ireland continues to have a somewhat fractured relationship with land and food production. Most recently, for instance, in the drive towards more factory-style production, many small farmers are losing out in the struggle to survive and could learn much from this Indian documentary, where such problems have been addressed.

One wonders, nevertheless, whether such evocative nature stories only speak to the already converted, while a majority audience, who appears at the outset to be non-environmentally driven, much less predisposed to this radical food and deep ecological philosophy and ethics of nature, is left simply accepting, if not embracing, the regressive status quo around cheap and easily available, albeit pesticide-ridden, food. One would hope not, but at least such evocative documentaries serve to play off these inherent contradictions and engage audiences in fruitful dialogue with various modalities, while celebrating food production and enticing representations of (wild) nature.

As cultural critics suggest, our representation of nature usually reveals as much, if not more, about our inner fears and desires as about the environment. Nevertheless, the two attributes can be regarded as coterminous, since our inner fears and desires often reflect, in large part, the "external" environment. For Thoreau and, later, Leopold, wilderness in particular was a state of mind as much as a description of a place. Both of these environmental activists championed a land alongside people ethic and not a land versus people ethic. Humans and, most specifically, farmers are a central environmental force shaping landscapes everywhere. *Timbaktu* directly calls to mind the provocative writings of Leopold, Thoreau, and others, while adding an indigenous Indian environmental layer of scholarship and engagement which Western academics and environmental documentaries across the world would do well to appreciate and emulate.

Timbaktu: A Textual Reading

Farming and food production—particularly concentrated on quality, security, and reliability, as well as being ethically produced—has become a newfound preoccupation across the world, including India, as an environmental reading of this documentary testifies. John Muir, the patron saint of wilderness, speaks of how the "earth hath no sorrow that earth cannot heal" and this truism serves as a recurring motif in this short documentary, alongside echoing the foundational environmental scholarship of Thoreau, Leopold, and others. Such a well-timed cautionary tale speaks to modern farming dilemmas around such difficult struggles as they fight against the pernicious drive to produce ever larger factory-style operations, rather than embracing more wholesome organic ones.

The documentary opens with what could be described as a conventional, romanticised, and clichéd image of birds singing, alongside a close-up shot of fruit and vegetables shown in their full glory. This display of an unashamed glorification of the bounty of nature is often dismissed by environmental communication scholars as signalling a so-called First-Wave essentialising of nature and pandering to a heightened romantic sensibility, which tends to reify nature and put it on a pedestal, as many romantic poets and writers over the centuries have done. It has been sometimes suggested that ecofilm, through its powerful appeal to the raw beauty of nature, tends to universalise and dismiss more critical injunctions and barometers of power and injustice. The drive to simply celebrate nature unconditionally has been considered as myopic with regard to a more critical engagement with nature, which is not something outside of humans but essentially part of our nature.

I would contend that audiences continue to need all types of celebratory engagement with nature—not just the shrill and sometimes preachy tombs of *Food Inc.* (2008) or *Supersize Me*, but also a more gentle and evocative style, such as that of *Timbaktu*, which evokes a deep, even spiritual, need for close communication with nature. However, through constant use of contextual and political rhetoric and discursive dialogue, the text sets up a range of incisive interventions. This trajectory is anchored by careful use of close-ups, including water, leaves, and hands clasping precious plants, or loose soil sculpted into them. Particularly resonant are the various evocations of a solitary tree framed in the landscape and the ethnographic purity of a gardener showing off his mango tree, while speaking of the power of natural compost from the leaves lying on the ground. Such close-ups are followed by an extreme close-up of seeds to illustrate how "we have lost our food security." But such pessimism around globalised dominance is wiped away by the beauty of "hard labour" with shots of bare muddy feet and oxen ploughing the land—no invasive factory farming machinery is witnessed—affirming the need to be in tune with nature, recalling many historical and seminal nature documentaries.

Countering these measured and carefully modulated close-ups are numerous mid-shots of precise landscapes that are carefully framed, including shots of the aforementioned mango tree. All the while, the narrator and the focus of the documentary define agriculture as the art of living with the land and nature and making the land feel good. Farmers and workers are constantly shown enjoying their labour in mid-shot, framed in total harmony with their habitat and leaving space for all to breathe. To link so many issues together, various blank screens are used, and transition shots help keep the narrative flowing along. Probably the image that stands out the most is an off-kilter iconic shot of a tree in mid-range, with the landscape flowing on either side, followed by a man in a white short loincloth standing proudly—like so many other workers in this documentary—in his green fertile field. Such photography dramatises the holistic vision of total synergy between man and nature, much beloved by the theorists of deep ecology mentioned above.

Such an apparently simple treatise around land husbandry continues, citing the respect the people have for the land by calling her Mother, nurturing her, and giving her a place of pride in their lives. The documentary ends with the chief protagonist and narrator recalling the first time he sat on a bench in this very landscape and felt the total "magic" of the place. This small "visual essay" certainly resonates with this magical feeling, capturing the quiet beauty and majesty of the farming community and nurturing habitat, which has been radically reclaimed as a viable and economic unit.

Timbaktu is littered with subtitles, much beloved by polemical environmental documentaries, sometimes so much so that it is hard to read all the comments on the screen. Nevertheless, the contextual and pointed comments and statistics display some fascinating facts about food and ecology, especially with regard to India, beginning with the fact that every native will consume vegetables containing over 40 pesticides. This preoccupation around the dangers of pesticides remains a core theme throughout and sets up a debate between the inherent benefits of organic farming as opposed to mass-produced, factory farming. For instance, we are provided with a practical illustration which serves to promote organic pesticides, costing only one dollar per unit, compared with 18 dollars for the artificial variety. Use of artificial pesticides has been abused to promote a mono-agricultural food system, while forcing the earth to artificially develop increased productivity. Such an artificial system of agriculture ends up with farmers spending more and risking their future survival and livelihood, as food security and human health become major issues. The spectre of a high rate of suicide is evoked to help dramatise the inherent dangers of depending on a precarious cash-crop form of agriculture.

The region chosen for this "back to nature" commune is a drought-prone plain in the Anantapur district of Andhra Pradesh region, where 7000 trees have been planted as a starting measure to help secure long-term and sustainable agriculture. The apparent leader of the commune and the focus of the documentary, C.K. Ganguly, remains a very engaging and charismatic figure for the movement. He initially humbly admits to knowing nothing about soil fertility or biodiversity. Such scientific knowledge remains a key marker of good farming and land husbandry, which in turn is necessary to ensure the earth remains fruitful and helps fight against the dangers of overproduction.

Recalling earlier environmental scholars who have been concerned about the dangers of simply anthropomorphising the land and nature generally, I would also contend that the language and philosophy evident in this documentary consciously evokes a deep, spiritual, and much-needed, interconnected celebration of agriculture, which not only is appealing to committed environmentalists, but also speaks to people from all walks of life. For instance, beginning with clearly unambiguous assertions, such as "land needs to be healed," it then shows close-ups of its concrete-like soil that needs careful nurturing. Good farming ostensibly encourages the earth to do her job. All "we" have to do is stop the burning of the land, using an aesthetic form of direct address and verbal rhetoric which remains both effective and provocative throughout. Consequently, the chosen area has grown from sustaining 21 species to over 400, all flourishing within this healing and nurturing habitat. Agriculture is the art of living with land and nature, as Thoreau and Leopold have affirmed. Make land feel good and give it care remains the underpinning mantra of the documentary as it sets out the essence of fruitful nurturing, which in turn embodies and almost defines good sustainable farming.

"A seed is God's gift to us" is pronounced like an Old Testament sacred edict and might seem out of place in a post-religious Western sensibility, but not here. Such religious and transcendental enlightenment philosophy might *not* go down well in the "scientifically driven" West, which denigrates any notion of a farmer's intuition or craft, much less talk of "God's gift." But this God–seed connection ought to be discussed more often in order to emphasise the long-term connection of man and nature within clearly coded environmental narratives. This connection is also where indigenous environmental philosophy and religious ideals become an issue and they should also be used to address Western environmental sensibilities. For example, Nirmal Selvamony and Rayson K. Alex talk of "*tiNai* criticism," based on the Tamil Dravidian concept of *tiNai*. According to Richard Kerridge, in a forward to *Culture and Media*, this is "the founding principle of pre-caste society in Tamil Nadu, defined by Selvamony as an 'organism-like society,' constituting a 'ground of existence' that unites human and non-human elements, culture and nature. This conception of relationship to land and all forms of life in that land 'is the one that makes land scared, in the *tiNai* sense of the term used by Alex'" (xv). Such a concept has importance for global engagement with a new form of land ethics, taking into account a more nuanced communal form of organic farming.

Currently, however, we are reliably informed that three giant corporations own over 50 % of all seeds within agriculture and, unfortunately, the small farmers have no control over the process. "We will have to eat whatever is thrown at us," announces Ganguly. Yet in India, only 3% of the food is processed, compared with approximately 80% in America, which, counter-intuitively for some, makes India seem "backward." But by any measure, these figures suggest the total opposite with regard to food quality and sustainability, in contrast with more dominant Western ideologies around processed food production methods. This fully fleshed-out food commune has developed an alternative economic system and a sophisticated social structure, as well as an educational structure. We watch adult trainee farmers learning about radically new methods of land husbandry in the fields, where they can observe their crops. These systemic changes portray a new organic and holistic society, sustaining a growing number of villages, communities, and families, which will hopefully withstand the ongoing commercial pressures to conform and adapt to the all but pernicious factory farming, cash-crop model, which is so anti-environmental and unsustainable in almost every way imaginable.

Conclusion

The documentary *Timbaktu* gently but persuasively illustrates the benefits of creating a farmers' co-op with memorable and beautiful scenes of farmer and workers in harmony with nature, including what might sometimes be considered as clichéd nature images of sunsets, alongside therapeutic sounds of bird song. Farmers as "nature citizens" are shown as essentially nurturing and artisanal in their approach to their duties, while working gracefully and harmoniously to complete their physical work. But to successfully produce food organically by using no artificial chemicals, prices have to be a little higher. Otherwise, who bears the risk of spoiled food, which is not an issue for processed and factory farming food systems, which refrigerates and uses harmful chemicals to grow and extend the life of their produce. Consequently, the documentary and its spokesperson call for more governmental protection and financial support for such healthy, environmentally sustainable, agricultural initiatives.

Timbaktu means "lost horizon," where seed meets the land. Reclaiming "waste" land remains essential for the overall food security of our planet, especially in highly populous areas like India. This documentary dramatises a most inspiring example of this necessary transformation. The audience is encouraged to look for this connection in their heart and to go deep within themselves. "Nature is part of us, if we seek to find it!" Such assertions evoke a romantic expression of deep ecology, as codified in the evolving eco-media literature. The documentary highlights how when local farmers have this feeling of co-ownership with the land, they can then effectively connect with the bigger symbiotic picture of humans functioning communally and effectively within a cohesive and holistic habitat. A state of connection with nature and the land echoes and remains a core tenet of deep ecological philosophy. This beautiful and evocative documentary ends with a close-up of snails slowly moving across a small patch of land. The message is clear: farmers, as environmental citizens, can and should actively love nature and protect their habitat and live with it, rather than abuse it for the cheap fix of cash crops that need expensive and dangerous pesticides to sustain growth.

More scholarly work is needed, however, to tease out the various alternative models and representational antidotes necessary to encourage more citizens to embrace this farming lifestyle as a bulwark to the pernicious excesses of economic growth that dominate global food production and agricultural methods generally. We also need more ecodocumentaries, such as *Timbaktu*, which address and hopefully assist in the process of gently changing behaviour patterns with regard to food production, alongside other related environmental concerns, before it is too late. Social sciences, alongside the humanities and media analysis in particular, in examining provocative documentary narratives such as this one, have a major part to play in developing fruitful, proactive, environmental discussions and in proposing and championing alternative and sustainable solutions to our farming and food demands.

I would finally endorse the widespread need for developing cross-cultural, environmental, literacy projects in order to be able to learn from such beautiful Indian case studies and *tiNai* principles cited earlier. Sustainability and environmental education essentially promote green cultural citizenship, which means embracing and embodying sustainable behaviours and cultural practices that shape and promote ecological values within the interconnected realms of society and the environment. As environmental literacy educators, we have a responsibility to address concerns around food quality and security in all its guises.

Notes

1. In holism, there is no nature/culture dualism: humans and nature are part of the same organic cosmological system. Theoretical ecologism often focuses its research on natural areas removed from human impact, human (or political) and so on.
2. "In Marx's view, when we speak of the value of a commodity, we are referring to exchange value. Of course the key question remains, is Marx also treating nature itself like an object or a commodity?" (Gandhi 2008: 42).

3. Agriculture—ploughing the soil—disturbs the native forest or grassland that preceded it. In this sense, all agriculture harms the ecosystem on which agriculture is superimposed (Rolston, 1994: 229). Feehan and other agricultural historians constantly speak of the long-term consequences of the Great Famine on the Irish landscape and on farming in particular.
4. What Marx said of history can be said equally of the environment: "[P]eople make nature, but not under conditions of their own choosing." The Gaia hypothesis is now a familiar example of this green logic: "[C]onstrained to consume what humanity produces, nature as subject creates a history and a future in which humanity may have no part. Nature in this instance takes responsibility for itself, but not for us. Its intelligence and agency are inhuman" (Cubitt 2005: 142).
5. Scholars talk of the need for a new kind of "eater's ethic" and the need for a connection to where one's food comes from. As global Western-based food retailers, such as Wall-Mart, the pesticide and agribusiness players, such as Monsanto, and the fast food and junk food purveyors, such as McDonald's and PepsiCo, all try to fight against "good food" agendas, we all need to be aware of changing agendas and fight against the so-called "greenwashing" of food.
6. When hunger takes a literal rather than a metaphorical form, it propels actions that serve to define what it is to be human—or to be inhuman. Food and water scarcity leads to both brutality and kindness in science fiction films, such as *The Omega Man* (1971) and *Mad Max 2* (1982).

References

Alex, Rayson K., S. Susan Deborah and Sachindev P.S., eds. *Culture and Media: Ecocritical Explorations.* Newcastle: Cambridge Scholars, 2014. Print.

Attfield, R., and A. Belsey, eds. *Philosophy and the Natural Environment.* Royal Institute of Philosophy Supplements. Cambridge: Cambridge UP, 1994. Print.

Bauman, Zygmunt. *Wasted Lives: Modernity and its Outcasts.* Polity, 2004. Print.

Brereton, Pat. *Environmental Ethics and Film.* London: Routledge, 2016. Print.

Brereton, Pat. *Hollywood Utopia: Ecology in Contemporary American Cinema.* Bristol: Intellect Press, 2005. Print.

Crocker, David Alan, and Toby Linden, eds. *The Ethics of Consumption: The Good Life, Justice and Global Stewardship.* Lanham, MD: Rowman and Littlefield, 1989. Print.

Cubitt, Sean. *EcoMedia.* New York: Rodopoi, 2005. Print.

Dobrin, Sidney, and Sean Morey, eds. *EcoSee: Image, Rhetoric Nature.* Albany, NY: State University of New York, 2009. Print.

Foster, John Bellamy. "Marx's Theory of Metabolic Rift, Classic Foundations of Environmental Sociology." *American Journal of Sociology* 105.2 (Sep. 1999): 336–405. Print.

Foster, John Bellamy, Richard York, and Brett Clark. *The Ecological Rift: Capitalism's War on the Earth*. New York: Monthly Review Press, 2010. Print.

Gandhi, Devadatta. "The Limits and Promise of Environmental Ethics: Eco-Socialist Thought and Anthropocentrism's Virtue." *Environs Environmental Law and Policy Journal* 31.1 (2007): 35–58. Web. http://environs.law.ucdavis.edu/volumes/31/1/gandhi.pdf

Gottlieb, Robert, and Anupama Joshi. *Food Justice*. Cambridge, MA: MIT Press, 2010. Print.

Gustafsson, Tommy and Kaapa, Pietari, eds. *Translational Ecocinema: Film Culture in an Era of Ecological Transformation*. Bristol: Intellect, 2013. Print.

Ivakhiv, Adrian. *Ecologies of the Moving Image: Cinema, Affect, Nature*. Waterloo, CA: Wilfrid Laurier UP, 2013. Print.

Ivakhiv, Adrian. "Green film criticism and its futures." *Interdisciplinary Studies in Literature and Environment* 15.2 (2008): 1–28. Print.

Jamieson, Dale. *Ethics and the Environment: An Introduction*. Cambridge, UK: Cambridge UP, 2012. Print.

Kellert, Stephen, in Richard Knight and Suzanne Riedel, eds. *Aldo Leopold and the Ecological Conscience*. New York: Oxford UP, 2002. Print.

Leopold, A. *A Sand Country Almanac*. New York: Oxford UP, 1947.

MacDonald, Scott. *Avant-Garde Film: Motion Studies*. Cambridge, UK: Cambridge UP, 1993. Print.

Merchant, Carolyn. *Radical Ecology: The Search for a Living World*. New York: Routledge, 1992. Print.

Murray, Robin, and Joseph Heumann. "Contemporary Eco-Food Films: The Documentary Tradition." *Studies in Documentary Film* 6.1 (2012): 43–59. Print.

Retzinger, Jean P. "Speculative Visions and Imaginary Meals: Food and the Environment in (post-apocalyptic) Science Fiction Films." *Cultural Studies* 22.3–4 (2008): 369–90. Print.

Rolston, Holmes. *Conserving Natural Value*. New York: Columbia UP, 1994. Print.

Rust, Stephen, Salma Monani, and Sean Cubitt, eds. *Ecocinema Theory and Practice*. London; New York: Routledge, 2013. Print.

Selheim, Megan Elizabeth. "Towards a Political Economy of Activist Documentary." Dissertation, Montana State University, 2011. Web.

Tong, Chris. "Ecocinema for All: Reassembling the Audience." *Interactions: Studies in Communications and Culture* 4.2 (2013): 113–28. Print.

Weik von Mossner, Alexa, ed. *Moving Environments: Affect, Emotion and Film*. Waterloo, CA: Wilfrid Laurier UP, 2014. Print.

Willoquet-Marcondi, Paula, ed. *Framing the World: Explorations in Eco-Criticism and Film*. Charlottesville, VA: U of Virginia, 2010. Print.

Wilson, E.O. *Biophilia*. Harvard: Harvard UP, 1984. Print.

PART II

Development, Waste, Ethics

Ecological Imperialism in the Age of the Posthuman: David Fedele's *E-Wasteland*

Başak Ağın Dönmez

David Fedele's 20-minute documentary *E-Wasteland* (2012) focuses on a serious environmental crisis emerging out of the heart of Africa. The documentary is sharp in its tone and direct in its message, and it also leads the audience to consider the highly possible future risks involved with this crisis. The electronic debris of the developed countries ends up in the economically disadvantaged regions of the world, where deprived people burn these overused technological devices to obtain copper in return for economic gain. The documentary displays the heartbreaking consequences of this practice. While the monetary advantage derived from the deadly experience of open-burning cables is apparently little and clearly insufficient for survival, the hazards caused by the detriments are incalculable. On the one hand, there are clearly visible effects, such as the infiltration of the molten objects into nature. On the other hand, some effects will most probably only become manifest in the long run, on both the flora and the fauna, as well as on people, for instance, in the form of fatal diseases caused by the inhalation of poisonous gas emissions. Almost a hundred years after T.S. Eliot's *The Wasteland* (1922),

---

B.A. Donmez (✉)
Middle East Technical University, School of Foreign Languages, Department of Basic English, Universiteler Mah., Dumlupinar Bulvari, 106800 Ankara, Turkey

R.K. Alex, S.S. Deborah (eds.), *Ecodocumentaries*,
DOI 10.1057/978-1-137-56224-1_5

the "unreal city" (60) becomes a reality in the Agbogbloshie slum in Accra, the capital of Ghana. "The brown fog" (Eliot, *The Wasteland* 61) is now replaced with the toxic fumes released from plastics, chemicals, and heavy metals such as lead, mercury, and cadmium. The e-waste of the First World, sent on a regular basis to the Third World for recycling, causes ongoing devastation in the more-than-human world. This is both environmentally and ethically troubling as the land, the air, and the water are polluted, causing harm to the human and the non-human alike, leaving the planet with irreversible scars. What Alfred Crosby coined as "ecological imperialism" in 1986 now takes a relentless form of neoimperialism. Crosby accounted for the European military expansion into the new worlds, maintaining that this was the obvious result of the technological superiority of firearms over relatively nascent forms of weaponry, such as spears. However, behind the displacement and replacement of the native peoples in temperate zones, he found another explanation, which could only be attributed to biological imperialism, rather than military success. The rationale behind the practice of displacement and replacement was no different from exploitation. "The plants on which Europeans have depended for food and fiber, and the animals on which they depended for food, fiber, power, leather, bone, and manure," he wrote, "tend to prosper in warm-to-cool climates with annual precipitation of 50 to 100 centimeters" (6). Evidently, the European settlers (or powerholders in general) wanted the most advantageous areas of the planet to be theirs. Thus, in the broadest sense, Crosby's explanation points out the reasons why the economically disadvantaged regions of the world are still suffering from poverty. This is not happening because they are at a disadvantage due to geographical and climatic causes, but because they have been ecologically exploited and deprived of the resources they once possessed. (The concept of "natural resources" itself is also problematic, and this will be briefly discussed at a later point). In Fedele's documentary, the case is not far from ecological imperialism, but it appears in its twenty-first-century form. The debris resulting from the technology produced in the First World would definitely cause damage to both human and non-human health, so it is dumped over the Third World instead. After all, why would the Third World matter? But in the lines that follow, it is clear that it *does* matter, not only in the political or ethical sense, but also in a posthumanist sense of *mattering*.

What Is Mattering? The Posthumanist Background to the Analysis of *E-Wasteland*

In Karen Barad's agential realistic account of the phenomena, which is a vital component of the posthumanist set of theories in its current formulation, "questions of ethics and of justice are always already threaded through the very fabric of the world" (Barad qtd. in Dolphijn and van der Tuin 69). In the Baradian posthumanist perspective, knowing, being, and valuing do not exist independently from one another. They are inseparable since "matters of fact, matters of concern, and matters of care are shot through with one another" (ibid.). All human and non-human bodies, organic and inorganic, biotic and abiotic, as well as the born and the made, are inextricably bound with one another and with the environment that is around and within them. As Barad also maintains, "the very nature of materiality itself is an entanglement" (ibid.). Inevitably, every decision made and put into practice in the First World would experience its consequences elsewhere, and vice versa. There is no escape, nor is there any possibility of "throwing" waste "away" in nature. It returns and strikes back. Coping with this necessitates a posthumanist understanding of eth ics, which requires that we rethink our sense of responsibility:

> Ethics is [. . .] not about right responses to a radically exteriorized other, but about responsibility and accountability for the lively relationalities of becoming, of which we are a part. Ethics is about mattering, about taking account of the entangled materializations of which we are part, including new configurations, new subjectivities, new possibilities. Even the smallest cuts matter. Responsibility, then, is a matter of the ability to respond. Listening for the response of the other and an obligation to be responsive to the other, who is not entirely separate from what we call the self. This way of thinking ontology, epistemology, and ethics together makes for a world that is always already an ethical matter. (Barad qtd. in Dolphijn and van der Tuin 69)

As such, the Western traditional conceptualisation of matter as more inferior than meaning-making practices would always fall short in configuring a new environmentalist discourse. This is because matter acts and triggers change in ways that we might not always fully grasp. Matter is alive. It "feels, converses, suffers, desires, yearns and remembers" (Barad qtd. in Dolphijn and van der Tuin 59). The chemicals released through the burning of the silicon-based bodies in Accra, as shown in Fedele's documentary, are causing harm to both human and non-human health. Having such agentic capabilities, they are alive in the new materialist sense. They are changing their forms, reaching different parts of the world, and leaving territorial and psychological marks on all carbon-based bodies of the planet. It is through these marks that Fedele highlights the potential dangers. Thus, in the material ecocritical sense, these chemical agents hold narrative powers. In narrating the story of the earth and environmental change, they are making an effect, so they are living and active. They become united with other (organic and inorganic) bodies, remembering themselves to us in the form of poverty or disease. Hence, what is neglected in the First World's exploitative practices is a crucial point. The electronic waste of the developed world not only brings about fatal consequences for the underprivileged people of Africa, but will also unavoidably lead to a larger-scale catastrophe.

Undoubtedly, then, waste, pollution, or debris cannot be taken as passive and inert. They do not remain where they are since they hold as much agentic power as humans do. They move. Agency here is not synonymous with human intentionality, consciousness, or cognisance but denotes a sense of effect, or the capacity to change the course of events, in a form of causality that is not always linear. This is in fact an "assemblage" of effects, as Jane Bennett writes. The assemblages "resonate with and against their 'causes,' such that the impact of any added element [...] or set of elements [...] cannot be grasped at a glance" (Bennett 42). The agency of these added elements is "slowly brought to light as the assemblage stabilizes itself through the mutual accommodation of its heterogeneous components" (De Landa 144). Put differently, the world's agentic potentials "intra-act" with all organic and inorganic bodies (Barad, *Meeting* x). As the earth's powers, such as rivers, ocean waves, and winds, carry the waste and electronic remains elsewhere on the planet, the harm caused by such pollution will not only affect its primary location, but will also spread to the rest of the planet. This is how, for instance, the Great Pacific Garbage Patch has materialised. The ironically romanticised tone of the short film *The Ballad of the Plastic Bag* (2012), which recounts the story of this huge garbage patch, demonstrates how plastic as a non-biodegradable artefact can "reproduce" in a way that is not unique to organisms. As Stacy Alaimo underlines:

> [a]lthough [*The Ballad of the Plastic Bag*] doesn't portray the agencies of the bag as it releases toxins or clogs an animal's digestive track, the clever conceit

> of the plastic bag as a ramblin' man dramatizes the agency and "freedom" of this supposedly inanimate object, stressing that these flimsy things have gotten away from us—escaping human control. ("Oceanic Origins" 196)

Believing in the possibility of taking every natural component under human control, then, is only a fantasy of the Western discourse of scientific progress. Things that are often thought of as inanimate are actively engaging in the course of human history. However, this does not necessarily absolve humans from any responsibility. On the contrary, as Barad's remarks on ethics urge us to change our pathways, this requires a modification in our sense of responsibility. It is necessary that we accept, as Lawrence Buell suggests, that "human history is implicated in natural history" (6). To employ a Turkish idiom here would be appropriate to describe the fallacy of the West in dumping its waste over Africa: the First World is cutting the branch (of the tree) on which it is sitting. As the entire world is in the midst of such an ecological crisis, this chapter aims to present an ecological critique of the current situation in Accra by looking at Fedele's *E-Wasteland*. The documentary is analysed from a posthumanist perspective that accounts for the networks and zones of intersection between the human and the non-human. Posthumanism, in this regard, is used in its various meanings. First, it is used in a relatively popular context as it highlights the twenty-first-century craze of technology; hence the title. Second, as presented above in the Baradian understanding of posthumanist ethics, it is employed as a new set of theories that defy the segregation of the human from the non-human as far as the agentic capabilities are concerned. In this sense, it is used in a new materialist form, in which non-humans share the agentic powers with their human counterparts. Matter in this new materialist sense has the capacity to act and trigger effect. Such an understanding of posthumanism derives its energy from an alternative approach to poststructuralism by putting matter and materiality back into the equation. Third, posthumanism is grounded on a political and philosophical plane as a marker of the expiration of liberal humanism. In this, it will also serve as a template for challenging the anthropocentric mindset of the West by deconstructing the major dichotomies between the centre and the margin, as well as the human and the non-human. Finally, in certain cases, a derogative meaning of the term "posthuman" (as a dystopian term that might also denote the end of humanity as we know it) is employed. As all of these uses appear several times in the chapter, I believe contextual and parenthetical explanations will clarify the framework in which the term is used.

Critical Questions on the Nature of Nature

The traditional Western ontological divide between nature and culture—as well as the human and the non-human—blinds our view, disabling us from being able to relocate ontologies, epistemologies, and ethics. A direct result of this inability to see ourselves as part of nature disconnects our bonds not only with our own bodies but also with the body of the earth. Indeed, all planetary bodies including all human and non-human ones, as well as that of the earth itself, are becoming contaminated, and that presents fundamental problems on political, ethical, and bio-environmental grounds. Fedele's documentary, therefore, echoes Karen Barad's posthumanist remark that "*we are a part of that nature that we seek to understand*" (*Meeting* 26; emphasis in the original). This is not, however, to suggest that we must abandon our pursuit of science and turn to seek a form of pristine nature or wilderness that needs to be preserved. On the contrary, such a suggestion would not only romanticise or idealise our relationship with nature but, even worse, it would reduce nature to a passivised object by reiterating an anthropocentric discourse. By neglecting the "dark side" of nature that can be devastating or frightening, we are fetishising nature and putting it on a pedestal. But this is not realistic. There are several questions that would follow such a glamourised perception of nature. Why, to begin with, should we think of nature as something far away from humans so that we must "go back" to it? Where and when did we lose our connection with nature? When did we start seeing ourselves as completely distinct entities from the environment that surrounds us? Why does technology, and its impacts, have to leave unalterable impairments on the planet? Recycling, yes, but at whose expense? And why would subjectivity belong to the human domain only? Does nature exist only at the service of humankind? Is nature merely "a resource" for human needs? Do we get the idea that we can "tame" or "dominate" nature simply because we belong to the species called *Homo sapiens*, which happens to possess a highly developed cortex? While these questions are obviously rhetorical and intended as a critique of the current forms of exploitative practices, there is one real question that might literally demand an answer. If nature is objectified and exploited in a similar way to how colonised people once were, we must ask Gayatri Spivak's famous question by reversing it: "Can the subaltern [not] speak?" (271). Indeed, the recent trends in ecocriticism and related philosophies have indicated that nature is eloquent. Basing their grounds on posthumanist theories in the new materialist sense, material ecocritics Serenella Iovino and Serpil Oppermann, for instance, argue that matter does not only possess agency in a manner that triggers effect, but also has the capability of generating meanings. They assert that aside from its impact-creating abilities, matter has the capacity of telling stories. As the host of countless organic and inorganic material bodies, therefore, nature cannot remain silent but is, in fact, story laden with its many components. For example, the geological strata within the lithic bodies of the earth tell us stories about the history of human and nature relations, as demonstrated in Jeffrey J. Cohen's *Stone: An Ecology of the Inhuman* (2015). Cohen maintains that the agentic powers of the stone have come to alter "cognitions, social constructions, scientific practices, and ethical attitudes" (11). One of his many poetical examples is as follows:

> A stone is that mundane object on which a philosopher might perch in order to think, ideation's unthought support; or in the palm, a spur to affect, cognition, and contemplation. Foundation of the inhabited world and its most durable affordance, stone is the material of our earliest tools, a lasting substance for our architectures, and intellectual ally. (Cohen 11)

Likewise, Iovino and Oppermann draw attention to the enmeshment of the human within the stories of the body of the earth. "The transformative stories built by telluric powers, magnetic forces, clashing and melting elements, and dawning forms of life," they point out, "extend the past of the earth into our present" (7). It is then clear that the human is not the only speaking subject and, yes, nature as the subaltern can speak. This view, although not on the literal level, is also configured as a confederate of agencies in Fedele's documentary as the documentary itself, being an inorganic body, speaks volumes in just 20 minutes. Moreover, the enmeshed status of the elements at work is telling us the stories of the future that will involve human suffering while the threat to our planetary existence grows. On an affirmative level, this echoes Christopher Manes' observation on non-Western traditions, in which nature is always thought of as capable of speaking, that is, *natura loquens*: "not just people, but also animals, plants, and even 'inert' entities such as stones and rivers are perceived as being articulate and at times intelligible subjects, able to communicate and interact with humans for good or ill" (15). Such an assessment, however, has two consequences: first, the physical marks that we leave on human and non-human bodies via our cultural practices will definitely tell sad stories about the history of the earth, and second, if humans are not the only speaking, rational, and cognisant agents of the earth, there evidently lies a problem in the boundaries drawn between the human and the non-human domains.

Who Counts as Human? Posthumanism as an Alternative Approach to Human–Nature Relations

Clearly, another follow-up question to the discussions and observations presented above would be "who counts as human?" If the human is defined as the centre of the discursive practices, then who are those others that are pushed to the margins and exploited? Can we not consider them as humans? There are clearly social, cultural, economic, political, and technological borders between the First World and the rest of the world, as also observed in Fedele's documentary. What counts as "scientific progress" belongs to those who are "developed," and what is considered to be "the rest" is always condemned to suffering, pain, and disease brought about by any form of exploitation, be it social, cultural, economic, political, or technological. And it is often the case that all forms of exploitation exist together and support one another in justification. Still, if technology is indispensable to the evolution of *Homo sapiens*, its impacts should be reconsidered on a global scale. Our cultural practices regarding our relations with nature call for a more horizontal alignment of the human and the non-human. This is not only necessary for a distributive justice for the world's human inhabitants, but also essential to understanding the agentic powers of the non-human, including those of impersonal agents like chemical bodies and plastic artefacts. Only with such a holistic, posthumanist approach can we continue theorising about any *eco*-movement. (Here, clearly the posthumanist approach refers to the one which aims at destabilising the borders between the human and the non-human.) Obviously, then, the answer lies in overcoming our human-centred mindset and in disabling the exclusionary definition of the human. The so-called universal human identity that is extremely generalised within the Western tradition under liberal humanist discourse is also problematised in several accounts by Spivak. "The great doctrines of identity of the ethical universal, in terms of which liberalism thought out its ethical programmes," Spivak contends, "played history false, because the identity was disengaged in terms of who was and who was not human" ("Remembering" 227). Drawing upon

Spivak's argument, Cary Wolfe also builds an analogy between speciesism, racism, and sexism. After all, humanism, as a "species-specific" discourse that strictly underlines the distinctions between the human and the non-human, has been extensively employed to "oppress both human and nonhuman others" (Wolfe, *Critical* 42). In its holistic but deconstructive strategies, posthumanism as a follow-up movement to poststructuralism offers an alternative and radical critique of human exceptionalism. From philosophical and environmental aspects, posthumanism puts a persistent emphasis on the breakdown of the boundaries between the human and the non-human. Wolfe, for example, argues:

> As long as this humanist and speciesist structure of subjectivization remains intact, and as long as it is institutionally taken for granted that it is all right to systematically exploit and kill nonhuman animals simply because of their species, then the humanist discourse of species will always be available for use by some humans against other humans as well, to countenance violence against the social other of whatever species—or gender, or race, or class, or sexual difference. (*Animal* 8)

Thus, all binary *-isms* co-constitute what counts as human and what does not. On the philosophical scale, this is a socially constructed ontological problem, which is sustained by our current epistemological methodologies and which supports our failing ethical approaches. As Bruno Latour also criticises the construction of "two entirely distinct ontological zones: that of human beings on one hand; that of nonhumans on the other" (10–11), posthumanism in a new materialist sense can offer a thorough alteration to such epistemological, ontological, and ethical problems by deconstructing the great divides between matter and meaning, nature and culture, and human and non-human.

The Nature/Culture Divide at Work

Traditionally, culture has been considered a social entity, while nature is taken as a bio-physical entity. [Hu]Man and *his* culture have been attributed a kind of power and a so-called ability to manipulate nature—as if the human species exists outside nature. Thus, any person/thing that has been assigned the role of superiority, such as the white, the male, and the Westerner, is "qualified" to be associated with such an understanding of culture. On the other hand, the non-white, the female, the queer, and the indigenous have been considered to be aligned with nature, as if they exist outside this culture. The problematic construction of such a dichotomy is highlighted by various scholars working within the environmental humanities, such as Val Plumwood. "Dualistic constructions of reason and nature, mind and body, spirit and flesh," she maintains, "create polarising metaphors and understandings of these elements which are woven through many kinds of social division in the dominant culture. These constructions erase the agency and contributions of women, the body, materiality and more-than-human world" (19). While Plumwood continues her argument by highlighting "the inability to see humans as ecological and embodied beings that permeates [W]estern culture" (19), the seeming impasse, presented by one's proximity to and distance from nature, has been deployed to inferiorise those who do not "fit in" the categories of the so-called superiority. Western discourse, essentially underlining the so-called divide between the humans and the Others, is actually based on this dilemma. Anyone who falls into the scope of Otherness, such as people of colour, women, queer people, and non-human animals, are to be associated with nature when they are seen as uncivilised, and on the other hand, they are blamed for being "unnatural," for what is cultural is actually naturalised in the Western discourse. According to Stacy Alaimo:

> [N]ature, as a philosophical concept, a potent ideological node, and a cultural repository of norms and moralism, has long been waged against women, people of color, indigenous peoples, queers, and the lower classes. Paradoxically, women, the working class, tribal peoples, and people of color have been denigrated because of their supposed "proximity" to nature, even as queers have been castigated for being "unnatural." ("Trans-Corporeal Feminisms" 239)

Apparently, the nature/culture divide is strongly manifest in the Western tradition and heavily deployed to battle against those who do not conform to the "naturalised" norms of the West. Fundamentally, the problem with this tradition of humanism is that it offers a so-called modernity that seems to allow hybridities and differences to exist and flourish, on the one hand, and on the other, it bases its assumptions on an essentialist view that considers a universal mode of humanity to be possible. Within the culturally constructed but "naturalised" norms of modernity, where a white-male-centred discourse is predominantly employed to suppress all the differences, polarised divisions are woven within and outside the categories of human culture. That is, the human is thought to be the centre of a non-human environment surrounding it, while this non-human environment is seen to exist only to serve the needs of the human. This Western humanist tradition, long holding the belief that there is an essential notion of identity that is universal, takes this so-called universal human essence as a central landmark, and it has been, in Cary Wolfe's words, deployed strategically "against other human beings for the purposes of oppression" (*Critical* 41). In other words, the so-called universal model for the human being is restricted to a definition provided by the white-male-centred discourse of the West. Accordingly, this universal human is actually the "saviour" of nature as a "damsel in distress," and it is the "master of the civilisation" for those who cannot "benefit" enough from this "high culture." Therefore, assuming the role of the provider of opportunities, this white male must bring opportunities to those who are deprived of them. As the master, *he* is actually benefiting from the workforce that is easily available, while "getting rid of" *his* own waste. Via this "species-specific" discourse called "humanism," in which the definition of the human is associated with what is civilised and modern, this so-called modernity creates a paradoxical approach. It both formulates "mixtures between entirely new types of beings, hybrids of nature and culture" (Latour 10) and suppresses the hybrids by configuring itself around monolithic centres. The construction of the social divisions of the human and the non-human spheres as distinct entities is the core of the Western argument of humanism, which attributes agency only to the white, male human, paving the way for neoimperialist and environmentally oriented strategies for exploitation of labour, of women, of people of colour, and of the non-human agents of nature, on the grounds that they have no contribution at all to the becoming of the world.

How "Naturecultures" Function in *E-Wasteland*

The main problem in question in Accra is that while the developed world provides us with the knowledge that exposure to toxicity can cause acute damage to both humans and non-human animals, ironically enough, they seem to take this piece of information as a form of "situated knowledge" (Haraway 183) and ignore the ontological relationality between the electronic debris, toxicity, and organic life. In his observational film, Fedele underlines this paradox and epitomises what can be called a posthuman dystopia, where toxic bodies emerge as threatening agents in the age of technological craze. Here, the word "posthuman" denotes the end of the human, or actually, the end of organic life as we know it. In this dystopia, the erasure of human existence from the planet relies heavily on the erasure of other species, while technology and detritus seem to take over. To use a different interpretation of the term "posthuman" in a new materialist sense, then, Fedele is actually underlining how both the material actors, like chemical agents, and the discursive practices of the West (on the great divides between nature and culture, or human and non-human) can prepare the ground for a world without organic life. Fedele, thus, presents an opportunity to the audience to view how "the material–semiotic actors," in Donna Haraway's terms (208), function in re-instantiating a formulation of imperialism in an ecological form. He does so through a display of the contradictory approaches of the West towards the environmental crisis. While the Western discourse marks the material bodies of the world as "resources" and justifies exploitation, the material bodies call for a rethink of the discursive practices. Matter, in its chemical forms in Accra, is reshaping and transforming the already fluid bodies of the planet, including the humans and the non-humans. This is, to borrow Barad's term once more, the "intra-action" of the bodies and discourses. The case of Accra is thus a live example of the material–discursive practices at work in the Baradian sense.

To clarify, Fedele's documentary highlights how matter and meaning are co-constituted and how the neocolonialist discourses of the West and related practices come into play in everyday life. On the one hand, the West discursively formulates an ontological divide between the human and the non-human, and it privileges what counts as "human" according to its own standards. On the other hand, feeding its own dichotomies of nature and culture, and privileging culture over nature, it dumps its waste into the lands of those who are underprivileged. Thus, Fedele's film successfully holds a mirror up to the key roles that the nature/culture divide and the actual inseparability of the two play. Accra, as presented in *E-Wasteland*, is just a microcosmic example of how the earth—as a non-human body—is systematically being corroded by humans. The film renders it possible to scrutinise the Third World's enmeshment of nature and culture, underlining the inseparability of human and non-human bodies. The human, the non-human animal, and the technological, all at once are at work as posthuman agents in an inextricable enmeshment. However, this posthumanity—because the epistemological and the ethical do not interplay in an affirmative mode—results in a dystopian reality on which the West turns a blind eye.

E-Wasteland, which shows Africa becoming the main place of disposal for cables, fans, and other countless types of mechanical parts, sent from the First World as electronic waste, opens with the statement that "every year, around 200,000 tonnes of second-hand and condemned electronic goods arrive in Ghana, West Africa" (00:02:53–00:03:03). The striking images of waste and hazardous industrial debris underscore how the luxury items of one part of the world turn into a death note for the rest of the planet. At the end of the film, the statistics are shown that "almost 50 million tonnes of e-waste are generated each year worldwide" (00:19:20–00:19:30), indicating the vast amount of waste that ends up in the Third World. The statements that these "goods are mainly received from the 'developed' world" (00:02:53–00:03:03) and that "a significant volume of electronics arrive as e-waste, exported illegally as 'second-hand goods'" (00:03:03–00:03:15) are actually the declaration of a hurtful truth. The Third World is now a site of toxicity, which accentuates the fact that environmental injustice goes hand in hand with social injustice. The documentary highlights the fact that people from the poorer regions of Ghana, who are generally uneducated, are drawn to the capital city of Accra in search of work, being forced to make a small living by recycling e-waste, the majority of which includes metals such as copper, aluminium, and brass. As these underpaid workers are open-burning electrical cables in order to salvage copper, in the background, amidst the toxic fumes, farm animals are wandering and searching for food, while children and women with babies are walking around in the hope of selling food and water. The imagery presented throughout the documentary is too harsh to be real. Life as we know it seems to be taken over by inorganic material, and the world looks as if it is experiencing its last days. Still, it is clear that in the philosophical posthumanist sense, the human and the non-human, the organic and the inorganic, the biotic and the abiotic are intermingled. This entanglement of matter and meaning is relocated on the physical and physiological levels, inflicting harm on the "naturecultures" (Haraway 11) of this posthuman embodiment. Such a "wasteland" of toxicity and industrial debris has apparently turned into a dismantled zone of corrosion. It is a vicious circle as new electronic waste keeps arriving, and the area is completely covered with all sorts of electronic, metal, and plastic material, from flat screens, computer monitors, and refrigerators to hair-dryers, tablet computers, and electric ovens. This toxic land of nature, permeated in its worst form by (human) culture, is not only an area of uncontrolled work conditions, involving underpaid adults threatened by job insecurity, but also an area of child labour: a young boy is shown trying to break a monitor into pieces with a stone. This is highly paradoxical, given the West's belief in the possibility of progress through science. The imposition of "scientific" management, as explained by Ted Steinberg, works on both the labour and the landscape level, through discourses of "controlling workers" and "controlling nature" (141). However, the belief that we need to tame nature and that we are capable of doing so is heavily problematic, as already discussed above.

Despite the expected fatal consequences of such unregulated working conditions, no precautions are taken to secure the health of the people of the region. The area, which is already heavily polluted by poisonous fumes and detrimental waste, is also the main living space for the citizens of the region, where they eat, sleep, and pursue leisure activities, such as playing board games and football. While the burned material is cooled down with water, making it even more convenient for the land to absorb its released toxicity, the West seems to prefer to ignore where its waste ends up, continuing to preach to the rest of the world about how to protect the environment and how to prevent environmental degradation. The ignored matter ends up intoxicating the finite "resources," including the human and the non-human. The final scenes of the film underline the fact that "recycling and burning e-waste releases toxic fumes from plastics, chemicals and heavy metals such as lead, mercury and cadmium, into the environment" (00:18:55–00:19:05). Yet, the people of the wasteland—partly because they are obliged to and partly because they are unaware of the dangers—along with the animals, are in direct and indirect physical contact with this toxicity as they touch the material, breathe the poisoned air, and drink from the polluted water. The amount of harm done to these people and animals is immense because, as the film notes, "exposure [to such toxicity] can cause acute damage to the lungs, blood cells and kidneys, increase the risk of cancers, and inhibit brain development, particularly in children" (00:19:06–00:19:19). What the First World does, as a result, is to inflict murder on the underprivileged inhabitants of the Third World and on the earth as a whole. On the continuum between providing labour opportunities and sacrificing the human and the non-human spheres, the First World's contradictory approach to toxicity is, thus, a result of the illusion of scientific management. This reminds one of John Rolfe's 1928 poem, "Asbestos," along with what Cary Nelson notes regarding this poem. As Rolfe writes: "John's deathbed is a curious affair: / the posts are made of bone, the spring of nerves, / the mattress bleeding flesh," Nelson comments that the "exploitation of workers… literally impresses itself on their bodies. Those bodies are the fulcrum, the point of application, of all the power relations in which their lives are embedded" (94). In a trans-corporeal reading of Rolfe's poem, Alaimo extends Nelson's interpretation and states that "the workers' bodies are not only the sites of direct application of power, but permeable sites that are forever transformed by the substances and forces—asbestos, coal, dust, radiation—that penetrate them" (*Bodily Natures* 30). From Nelson's and Alaimo's perspectives, it is possible to apply the workers' case in Rolfe's poem to the case of the workers in Accra and charge that the First World not only exploits the land as a disposal site, ignoring the extensive effects of this hazardous activity, but also employs the bodies of the workers in Accra in a twofold manner. This is the outcome of a profit-minded approach, which benefits from a cheap labour force, and this is also a penetration of the human body, along with which a number of non-human bodies are forever transformed. Culture, in its electronic, toxic, and plastic forms, infiltrates nature. Culture, in its power relations form, supersedes nature. Despite the argument of the inseparability of the two in a non-hierarchical manifest, the politically engaged cultural practices of the West heavily dominate what is "natural," if any naturalness remains. Fedele's documentary epitomises the intertwined framework of politics and the exploitation of nature(s), as the land, as the air, as the water, as the human, and as the non human bodies.

So far, this may sound as if the ideas presented here "subscribe to an ontology in which the material world is a passive resource for the exploits of the [Westerner] human" (Alaimo, *Bodily Natures* 31). However, what I am trying to clarify is the West's failure to see that there is no real distinction between culture and nature. Fedele's film is reflecting a state of ecological imperialism when it documents the West's use of the Third World as its garbage disposal area. This is actually an outcry for environmental justice, which, to quote Alaimo, "insists upon the material interconnections between specific bodies and specific places, especially peoples and areas that have been literally dumped upon" (*Bodily Natures* 28). This is the exemplification of a so-called environmental benefit gained by the West as it "gets rid of" its own garbage, while the human and non-human bodies of Accra are made to suffer from the environmental destruction, resulting from a dichotomy that separates nature from culture and the human from the non-human. Moreover, the privileged status of the mind, when compared with the body, retains its central position within the hands of the Western humanist discourse as materiality is often undervalued, if not totally disregarded. Despite the rise of environmentally oriented policies within the Western world, such as those concerning climate change and the protection of biodiversity, the underprivileged parts of the world are still neglected. As Fedele's film exemplifies, in the case of the Third World, the material agency of the inorganic and technological—like the metals and toxic gases that emerge as a result of the open burning of electronic waste—is not taken into consideration. The First World acts as if Ghana, West Africa, were a totally separate planet that did not have any effect on the First World's environment. This is either because the mind and meaning are still thought to be superior to matter, body, and materiality or because the peoples of the Third World are not counted as "human," or even worse, both. In Accra, the posthuman materialises both as the end of the human and as the indivisibility of matter and discourse. The dichotomies are torn apart as the humans lose their central position to the non-humans. Yet, this signals the end of history as we know it. There, in the capital of Ghana, electronic waste and toxicity have overthrown the so-called human masters.

In the light of what has been presented up to this point, what *E-Wasteland* portrays as a documentary film is clearly a social outcome of the Western policies on the Third World. It is, in fact, a sociopolitical, cultural, and ecological form of imperialism, in which it becomes more clearly visible that nature and culture are inseparable. Nature's construction as the surrounding environment of a human-centred world, which is full of resources to exploit and benefit from, is not only theoretically problematic, but also results in a failure to conduct healthy relations between humans and nature. As culture has been socially, historically, and politically constructed as superior to nature under the guise of human civilisation, the "developed" countries continue to justify their exploitation of underprivileged peoples and lands. What lies at the heart of the problem in Accra, therefore, is the Western idea that polarises socially constructed divisions as culture/nature, human/non-human, mind/body, white/non-white, and male/female. It makes it possible to exploit not only the labour force, but also the land and the resources of a less-developed part of the world. The polarised distinctions not only allow the West to label itself as the superior, but also illustrate its profoundly ambivalent relationship with nature, fluctuating between a romanticised commitment (to nature) as its protector or saviour and a careless and disinterested attitude that conquers or destroys it, not caring about what is left behind. In the case of Accra, both roles are intertwined. The West functions as the labour provider and the opportunity giver as well as the exploiter and the coloniser. The consequences of such an entanglement is a posthuman dystopian environment, which is corroded through an infiltration of culturally produced matter into nature's corporeality.

Conclusion

Obviously, the social divisions between nature and culture, mind and body, and the human and the non-human need to be re-questioned through an environmental and holistic frame. It must be acknowledged that nature does not hold an instrumental value, but an intrinsic value, while the human must be accepted as an ecological and embodied being, which is interdependent on others to survive. Nature is not a tool to be utilised, nor can it be tamed. The human is simply another biological species that is heavily dependent on the earth's life-support systems. Any view that fails to acknowledge this condition of the human is anthropocentric, and hence poses an obstacle to sustaining the balance of nature. Thus, from a posthumanist and an ecologically oriented perspective, it is possible to defy human exceptionalism as nature cannot be distinguished from culture, as Ylva Uggla argues, and nor can the human be separated from the non-human. For Uggla, "the notions of nature and the natural, as distinct from culture and society and untouched by humans can be questioned since we cannot find any site on earth that fits that description," and "the idea of nature and culture as separate entities is flawed" (80). The same argument is also shared by Glen A. Love and Lawrence Buell, who also join Plumwood's and Uggla's perspectives. While Love challenges "our notion that human beings are so special that the earth exists for our comfort and disposal alone" (229), Buell notes how significant it is to understand that "the nonhuman environment is present not merely as a framing device but as a presence" (6). Understanding the inseparability of nature and culture, thus, seems to lie at the heart of an ecological discourse in order to fight ecological imperialism. To highlight this indivisibility, Erik Swyngdouw has come up with an understanding of nature as "socio-environmental arrangements" (20) and, in a similar attempt, Matthew Gandy calls nature "seeming nature" (110). However, perhaps the best-known and the most widely acknowledged term has been coined by Haraway, as "naturecultures" (11), through which she highlights that these seemingly distinct entities are indeed united in a state of enmeshment. Elaborating on a similar mindset about the enmeshed framework of naturecultures, Marilyn Strathern notes that "there is no such thing as nature or culture, [but] each is a highly relativised concept whose ultimate significance must ultimately be derived from its place within a specific metaphysics," and she continues: "No single meaning can be given to nature or culture in Westernized thought; there is no consistent dichotomy, only a matrix of contrasts" (177).

As discussed above, from the traditional Western point of view, the potential agency of the non-human, the inorganic, and the abiotic is also disregarded, and what(ever) is associated with inferiority in the eyes of the West counts as "the non-human," being deprived of the agentic capability of creating effect and changing the course of events. On the other hand, in Fedele's film, it is clearly discernible that the peoples of Accra are also considered to be non-human, whose agency and existence are ignored by the First World. They are both economically and ecologically exploited, and their homeland is infected by the waste of the developed countries. Given such a paradoxical state attributed to the division between nature and culture in the Western tradition, and considering how the posthumanist way of thinking poses a challenge to this traditional set of values assigned to a certain "naturalised" culture, Fedele's film, *E-Wasteland*, brings up the quandary of human/non-human as the underprivileged people of Accra are forced to face inhumane conditions. At the same time, the film successfully presents the erosion of the boundaries between nature and culture as the cultural products of the West infiltrate the human and non-human bodies, turning them into toxic elements of naturecultures. To stop this catastrophe, of course, would require many more efforts, such as that of Fedele, as well as a call for a paradigm shift in both the material and the discursive senses.

References

Alaimo, Stacy. *Bodily Natures: Science, Environment, and the Material Self.* Bloomington: Indiana UP, 2010. Print.

Alaimo, Stacy. "Oceanic Origins, Plastic Activism, and New Materialism at Sea." *Material Ecocriticism* Eds. Serenella Iovino and Serpil Oppermann. Bloomington: Indiana University Press, 2014. 186–203.

Alaimo, Stacy. "Trans-Corporeal Feminisms and the Ethical Space of Nature." *Material Feminisms.* Ed. S. Alaimo and S. Hekman. Bloomington: Indiana UP, 2008. 237–64. Print.

Barad, Karen. *Meeting the Universe Halfway: Quantum Physics and the Entanglement of Matter and Meaning.* Durham & London: Duke UP, 2007. Print.

Bennett, Jane. *Vibrant Matter: A Political Ecology of Things.* Durham: Duke UP, 2010. Print.

Buell, Lawrence. *Environmental Imagination: Thoreau, Nature Writing, and the Formation of American Culture.* Cambridge: Belknap, 1995. Print.

Cohen, Jeffrey J. *Stone: An Ecology of the Inhuman.* Minneapolis: U of Minnesota, 2015. Print.

Crosby, Alfred W. *Ecological Imperialism: The Biological Expansion of Europe, 900–1900.* Cambridge: Cambridge UP, 2004. Print.

De Landa, Manuel. *Intensive Science and Virtual Philosophy.* London: Bloomsbury, 2002. Print.

Dolphijn, Rick, and Iris van der Tuin. "Matter Feels, Converses, Suffers, Desires, Yearns and Remembers: Interview with Karen Barad." *New Materialism: Interviews and Cartographies.* Ann Arbor, MI: Open Humanities, 2012. 48–70. Print.

Eliot, T.S. *The Wasteland.* 2004. 13 Feb. 2016. Web. <http://www.larecherche.it/public%5Cproposte%5CProposta_Poesia%5Cupload_pdf_doc_txt%5Cpaolom68_20151031093750_T.S.Eliot%20-%20The%20Wasteland.pdf>

E-Wasteland. Dir. David Fedele. 2012. Film.

Gandy, Matthew. *Concrete and Clay.* Cambridge: MIT Press, 2002. Print.

Glotfelty, Cheryll, and Harold Fromm, eds. *The Ecocriticism Reader: Landmarks in Literary Ecology.* Athens: U of Georgia, 1996. Print.

Haraway, Donna. *Simians, Cyborgs, and Women: The Reinvention of Nature.* New York: Routledge, 1991. Print.

Iovino, Serenella, and Serpil Oppermann. "Introduction: Stories Come to Matter." Serenella Iovino and Serpil Oppermann, eds. Material Ecocriticism. Bloomington: Indiana UP, 2014. Print.

Latour, Bruno. *We Have Never Been Modern.* Cambridge: Harvard UP, 1993. Print.

Love, Glen. "Revaluing Nature: Toward an Ecological Criticism." Cheryll Glotfelty and Harold Fromm, eds. *The Ecocriticism Reader. Landmarks in Literary Ecology.* Athens: U of Georgia P, 1996. 225–40. Print.

Manes, Christopher. "Nature and Silence." Cheryll Glotfelty and Harold Fromm, eds. *The Ecocriticism Reader. Landmarks in Literary Ecology.* Athens: U of Georgia P, 1996. 225–40. Print.

Nelson, Cary. *Revolutionary Memory: Recovering the Poetry of the American Left.* New York: Routledge, 2003. Print.

Plumwood, Val. *Environmental Culture: The Ecological Crisis of Reason.* New York: Routledge, 2002. Print.

Rolfe, John. "Asbestos." 1928. *Anthology of Modern American Poetry.* Ed. Cary Nelson. Oxford: Oxford UP, 2000. Print.

Spivak, Gayatri Chakravorthy. "Can the Subaltern Speak?" *Marxism and the Interpretation of Culture.* Ed. Cary Nelson and Lawrence Grossberg. Urbana: U of Illinois, 1988. 271–313. Print.

Spivak, Gayatri Chakravorthy. "Remembering the Limits: Difference, Identity and Practice." *Socialism and the Limits of Liberalism.* Ed. Peter Osborne. London: Verso, 1991. 227–39. Print.

Steinberg, Ted. *Down to Earth: Nature's Role in American History.* New York: Oxford UP, 2001. Print.

Strathern, Marilyn. "No Nature, No Culture: The Hagen Case." *Nature, Culture and Gender.* Ed. Carol P. MacCormack and Marilyn Strathern. Cambridge: Cambridge UP, 2001. 174–222. Print.

Swyngedouw, Erik. "Impossible 'Sustainability' and the Postpolitical Condition." *The Sustainable Development Paradox: Urban Political Economy in the United States and Europe*. Ed. Rob Krueger and David Gibbs. New York: The Guilford, 2007. 1340. Print.

Uggla, Ylva. "What is This Thing Called 'Natural'? The Nature-Culture Divide in Climate Change and Biodiversity Policy." *Journal of Political Ecology* 17 (2010): 7991. Web.

Wolfe, Cary. *Animal Rites: American Culture, the Discourse of Species, and Posthumanist Theory*. Chicago: U of Chicago, 2003. Print.

Wolfe, Cary. *Critical Environments: Postmodern Theory and the Pragmatics of the "Outside."* Minneapolis: U of Minnesota, 1998. Print.

Wasted Humans and Garbage Animals: Deadly Transcorporeality and Documentary Activism

Chia-ju Chang

Introduction

The 2014 tiNai Ecofilm Festival (TEFF) raised its curtain once again at the Birla Institute of Technology and Science, Pilani and Goa, India, on 30th January. Archiving a total of 43 films, the festival featured six countries, including Australia, Ireland, India, Portugal, the UK, and the USA, and showcased a variety of both local and global ecological issues from around the world. The repertoire included wildlife conservation and sustainability, indigenous ways of life and knowledge, commons/eco-village building, the impact of modernisation and transnational exploitation, food security and sovereignty, colonisation by foreign invasive species, animal protection, the waste problem, river pollution, climate change, and nuclear and other eco-activism/protests. Whether ostensibly activist, political, or more aesthetic and awe-inspiring, these films can be conceptually divided into both the deep and shallow ecology camps. While the former injects

---

I would like to express my utmost gratitude to the president of the Karuna Society, Clementien Pauws, and also to Kunal Vohra, the director of *The Plastic Cow*, for providing me with stills from the documentary and details relating to "The Plastic Cow Project."

C.-j. Chang (✉)
Department of Modern Languages and Literatures, Brooklyn College, 2900 Bedford Ave, New York, NY 11210, USA

R.K. Alex, S.S. Deborah (eds.), *Ecodocumentaries*,
DOI 10.1057/978-1-137-56224-1_6

hope and optimism by rediscovering enchanted nature and evoking a deep ecological sense of biophilia, nature's wonders, richness, and biodiversity, the latter turns its camera on the dark and "inconvenient" side of the earth's story. It focuses its gaze on the terrorism we wage against the land and oceans, as well as on group activism in fighting against the aggressive encroachment of capitalist modernity, and the corruption of politicians and corporations. Despite their differences, both camps share the same vision of preserving the ecological health, integrity, and sanctity of the earth—our only home.

Prof. Anjali Monteiro and Prof. K.P. Jayasankar rightfully commented on these films as making visible "marginal and subaltern ways of seeing and being that challenge the current growth driven dominant paradigm" (tiNai Ecofilm Festival Guide 2014). Here, the topic of waste, along with its related motifs of waste management, human body and pollution, dump-sites and recycling, environmental injustice, and so on, dovetails the rising material ecocriticism or, more specifically, what Susan Signe Morrison called "waste studies."[1] It also echoes a growing body of eco-documentary production worldwide that deals with the issue of waste: *Manufactured Landscapes* (Canada, 2007), *E-Wasteland* (Australia, 2012), *City's Edge* (India, 2012), *The Garbage Trap—Sanitary Workers of Indian Cities* (India, 2012), and *Beijing Besieged by Waste* (China, 2013), to name but a few. In this genre on waste, the themes range from large-scale landfills or waste-scape, waste management, toxic habitation, plastic pollution of the ocean, greenhouse gas emissions, and archaeological exploration of trash to existential–poetic or aesthetic musing on the lifecycle of a piece of garbage.

Part One: Waste and Modernity

The English word "waste," which initially referred to "a land or an environment that was unsuitable to sustain human habitation," also denotes the meaning of imbalance. In further elaborating on the connotation of imbalance, John Scanlan, the author of *On Garbage*, broadens this idea of imbalance to include "our contemporary relationship to places, objects, as well as to behaviors and practices" and, more specifically, narrows it down to a "perceived excess of consumption" (n.p.). These are examples of a modern way of life which dictates that the future fate of the planet is to run amok—to go out of control—as has already been famously exemplified in the documentary film *Koyaanisqatsi* (USA, 1982), with its Hopi title referring to "life out of balance" or "life of moral corruption and turmoil." Considered in this light, the discourse of waste conceptualised by the current hyper-consumerist mode of life out-of-control goes far beyond being a mere discussion on public health and environmental management and policy. Moreover, to frame the concept of waste as an imbalance from the perspective of global environmental justice, it is then interwoven into geopolitics relating to the imbalance of hemisphere/ continent, nation, race/ethnicity, class/caste, and gender. As shown in the festival, several translational eco-documentaries address the issue of exporting waste (and the unbalanced, exploitative nature of consumerist lifestyles) to developing countries or areas. This excessive (toxic and non-biodegradable) material waste has now emerged as a symbol of Western/ global modernity and environmental imperialism.

This vein of waste discourse needs to squarely confront a form of violence that is hidden behind the façade of the developed nations' practice of NIMBY (not-in-my-backyard) environmentalism. At the beginning of *Slow Violence and the Environmentalism of the Poor*, Rob Nixon quotes the confidential memo of the president of the World Bank, Lawrence Summers, who advocated exporting toxic waste and polluting industries to lesser developed countries in areas such as Africa.

Waste constitutes a form of Nixon's "slow violence," which, contrary to the instantaneous, spectacular, and visibly sensational "conventional violence," "occurs gradually and out of sight, a violence of decayed destruction that is dispersed across time and space, an attritional violence that is typically not viewed as violence at all" (2). This slow violence easily slips under the radar, but its calamitous repercussions tend to play out across a range of temporal and geographical scales. While the conventional cognition of violence is immediate and explosive, erupting into an instant of concentrated visibility and "newsworthy because it is focused around an event, bounded by time, and aimed at a specific body or bodies," (ibid.) the effects of slow violence are more pervasive but elusive. Examples of slow violence are "[c]limate change, the thawing cryosphere, toxic drift, deforestation, the radioactive aftermaths of wars, oil spills, acidifying oceans" (ibid.). Fittingly, as many scholars have pointed out, slow violence reminds one of Timothy Morton's hyperobjects. Due to the scale, on both macro- and micro-levels of time and space, which are beyond the normal sensory grasp, they mask violence. Sometimes, it takes hundreds of years to see the effect, or to legitimise it as a form of violence in order to take appropriate measures. It is hard for us to see the connection between two seemingly distinct things unless we see clear evidence of their linkage.

Taking racism and environmentalism as an example, there might be a deeper socio-psychological root in the practice of environmental racism. From a psychoanalytic viewpoint, waste, together with its cultural and material associations with unsustainability, imbalance, perverse decay, death, in-between, out of place, and ambiguous status, stands for something that is on the threshold of life and death, and has the potential to infect and threaten life with its impure content and associations; therefore, it presents a collective threat of what Scanlan called "civilisation's double," or what Julia Kristeva called "abjection." Abject, according to Kristeva, is "what disturbs identity, system, order" (4). Barbara Creed maintains that "the body protects itself from bodily waste such as shit, blood, urine and pus by ejecting these things from the body just as it expels food that, for whatever reason, the subject finds loathsome" (9). So, in light of this psychoanalytic school of thinking, it is not hard to understand that human civilisation is established and maintained by keeping the waste at bay. The more "developed" and urbanised the society becomes, the more clearly the line has to be drawn so as to maintain order. Zygmunt Bauman, in *Wasted Lives: Modernity and Its Outcasts*, writes, "[w]e dispose of leftovers in the most radical and effective way: we make them invisible by not looking and unthinkable by not thinking" (27). For Bauman, "the survival of the modern form of life… depends on the dexterity and proficiency of garbage removal" (ibid.). He quotes an anecdote recounting how Ivan Klima having dinner with the president of the Ford Company. The guests asked the president how to remove the massive numbers of cars from the world once they were no longer wanted. With confidence, the president answered, "Anything that was manufactured could vanish without trace," (4) and for him it was just a technical matter. Later, Klima discovered that the "deceased" cars were squeezed by gigantic presses into neat metal boxes. "But those metal boxes did not vanish from the world…" (ibid.). The shadowy existence of modernity and the lingering residue of its obsessive capitalist consumption can never vanish without a trace. In the end, we will become one with what we discard, as we are producing it, living it, and eating it. Eventually, we turn into the very embodiment of the toxic monster we create, whatever it is.

Human as Waste

Like Nixon, both Bauman and Scanlan prompt us to think of global environmental injustice. If the waste product, sewage, and the dumpsite are the ultimate expressions of unholy abjection, then the people associated with waste clearance are naturally also associated with impurity and defilement. "[T]he development of civilization keeps us from encounters with the abject. Instead, we leave garbage to be dealt with by others" (Scanlan, n.p.). Similarly, Bauman writes, "the economic progress…is spreading to the most remote nooks and crannies of the 'filled up' planet, trampling on its way all remaining forms of life alternate to consumer society" (59). Countries such as China, India, Vietnam, and Africa, serve as the "recycling" sites of the electronic waste (e-waste) of the West. In this "production–consumption–recycling" loop of material exchange between these two worlds, refuse workers at the dumpsite in these countries are the victims of environmental injustice; they are treated just like the objects of their work. Scanlan writes, "[l]ike the wastes and mess they look after they are degraded because, according to Michael Walzer, their relationship to dirt, waste and garbage 'makes them the object of disdain and avoidance'" (n.p.). They have become doubly outcast, both within their own society as well as in the world's market economy; they are what Zygmunt Bauman calls "human waste" or "wasted humans" in the process of modernisation: "the 'excessive' and 'redundant,' that is the population of those who either could not or were not wished to be recognized or allowed to stay" (5).

E-Wasteland (2012) and *City's Edge* (2011) both examine the communities of "human waste" by focusing, respectively, on two mega-dumpsites on different continents. David Fedele's *E-Wasteland* is a one-man guerilla documentary set in the Agbogbloshie e-waste dump in Accra, the capital of Ghana. It is the largest e-waste dumpsite in West Africa. According to the official site of *E-Wasteland*, the e-waste recycling business jacks up the demand for second-hand electronic equipment in developing countries. Every year, around 200,000 tonnes of second-hand electrical goods are exported illegally from the "developed" world as second-hand goods in Ghana.[2] E-waste consists of numerous toxic materials such as arsenic, cadmium, lead, and mercury. Once the waste is broken down, these toxins contaminate the air, water, and food consumed in the slum.[3] The refuse workers, usually uneducated young men, do not have health or environmental protection.

Fedele chose, as his strategy to approach this subject matter, to use no dialogue or voiceover. Disagreeing with other filmmakers' use of interviews and voiceovers to portray these e-waste workers as victims, Fedele says, "I was extremely conscious that I didn't want to do this. I wanted to visually present a particular environment, attempt to show it as truthfully as possible, and give people the responsibility to think about the issues themselves" (ibid.). Trusting in the power of the image and the audience's ability to engage in active, reflective thinking about the condition of a contemporary lifestyle that is saturated with e-products, Fedele continues:

> I believe that images can be far more powerful than words, and can have a much bigger impact and remain in your memory longer…I want people to contemplate where "things" end up at the end of their life. For me, that is what the film is about—where THINGS go at the end of their life—[and] it is as much a film about "things" as about "people"…But I was also wanting to show that these THINGS have their life in the "developed world," come to Africa once they are junk to be broken back down into their "parts," and then exported back to the developed world to continue making electronics for the developed world to use and benefit from—while leaving behind illness and sickness, pollution, and environmental destruction. (ibid.)

The director adopts the idea of unwanted material having a lifeworld of its own and material agency finds its echo in the American-Iranian filmmaker Ramin Bahrani's *Plastic Bag* (2009). *Plastic Bag* creates a new filmic language examining material agency through the use of anthropomorphism, irony, and over-the-top sentimentalism to the point of ridiculousness so as to ask the audience to reflect on our short-lived "love affair" (obsession and betrayal) with plastic bags in the age of what I call "Plasticocene" (to create a more nuanced shade and vocabulary for the age of the Anthropocene to address the issue of plastic pollution). Here the subtext (the endurance of the substance) captured in the journey of this overly philosophised, immortal plastic bag (i.e., venturing into the ocean and a marine trash vortex) narrates a cautionary tale of environmental justice.

City's Edge is set in India's Deonar Dumping Ground, which is said to be the largest dumping ground in Asia. The film was part of a larger initiative, the "M-Ward" project, which aims to address the problems faced by the residents of the most impoverished ward in Mumbai. From the synopsis and the film, we learn that thousands of trucks dump half of all the garbage the city produces at the Deonar Dumping Ground every day. And most of it finds its way back, as it is picked and processed by the many men (but mostly women and children) who eke out a living as ragpickers. They live on the margins of the city at the dumpsite. As the synopsis points out: "[T]he dump is a powerful symbol of most of the city's people who are constantly dumped, recycled, and dumped again, while perpetually remaining out of sight."[4] Not being able to be assimilated into society, or to find a job, they again exemplify Bauman's "human waste."

While Fedele presents a "voice-less" strategy to enable the audience to concentrate on the visual aspect, *City's Edge* uses live conversations, title cards, and superimposed texts to provide background information as a way to raise awareness. This documentary employs no voiceover or background music. It allows the images to speak for themselves, with only intermittent English dialogue providing minimal information at the beginning and towards the end of the film, in order to highlight global environmental injustice. Moreover, the use of children as a subject in documentaries on toxic waste is a powerful eco-communication strategy as it tends to evoke strong moral sentiment. In *City's Edge*, we see youngsters working at the dumpsite and exposing themselves to toxic substances without any protection. The film centres on the stories of two characters: Mohammad Hussain (aka Babu), a young ragpicker, and Haroon, a local scrap dealer. From their perspectives, we see that the dumpsite serves not only as a workplace and dwelling place for the outcast refuse workers, but is also the children's playground, where they scavenge and play.

If waste is the symbol of imbalance, decay, chaos, and abjection, all of which is to be avoided, then the people who work in the area of waste management become tinged with similar associations. This explains why the undesired waste-scape represents the shame of urban development, which needs to be made invisible and forgotten so as to sustain the illusion of a consequence-free urban lifestyle. In the case of India, the caste system becomes enmeshed in the waste-scape. Here, waste (as "unwanted matter") is tightly linked to caste ideology on two levels: first, there is the notion of "outcaste" and, second, due to the outcaste status, people within this group have no choice but to take up filthy jobs, such as refuse or sewage collection. In return, their job defines their class and becomes their only identity or, vice versa, the class into which they are born defines their job choices. Though not a TEFF 2014 film, the documentary *The Garbage Trap—Sanitary Workers of Indian Cities* sheds light on the condition of sanitary workers (street sweepers, drainage workers, etc.) within the city and further problematises the issue of class/caste and waste management. It becomes obvious that waste has become a marker of class between those who create it and those who clean it (refuse workers, formerly called Bhangis, who carried the excrement). The Dalits are not considered "humans"; they do all the dirty work, including collecting the cuts of animals, intestines, and human corpses.

These above-mentioned eco-documentaries from different parts of the world exemplify slow violence against the vulnerable groups of people who eke out a living by scavenging parts from extremely toxic waste. Thus far, the discussion has been focused on the health of *Homo sapiens* and the environment and non-human animals have not been included in the equation, even though traces of them, as an essential part of the waste-scape and community, can be found in *E-Wasteland*, *City's Edge*, and *Plastic Bag*. I now propose to contribute to the three health issues (healthy economies, environments, and bodies) elaborated on by Heather I. Sullivan and Dana Phillips in their introduction to the special edition for the Interdisciplinary Studies in Literature and Environment (*ISLE*) on dirt theory and material ecocriticism. I will leave aside our awareness of our own personal ecosystems, the matter of living organisms "on your skin, under your shoes, in your digestive tract, and in your very cells" (445), to focus on the bodies of animals. In developing a zoocentric, ecocritical, critical language discourse, and a praxis that fuses waste studies with critical human–animal studies, my aim is to unveil stories of slow violence against non-human animals, against their fur/feathers, their hoofs/horns, and their digestive tracts as well as their very flesh.

Part Two: "Animal as Waste" Vocabulary List: Trash Animal

When the global mainstream media discusses non-human animals in the context of environmental degradation, pollution, or disasters, they tend to take up an anthropocentric/speciesist position that focuses on the impact of non-human species on the interests of humanity (e.g., the disappearance of bees, SARS, mad cow disease, AIDS, etc.).[5] Taking the US media as an example, a recent CNN headline news item, "Unfit for Human Consumption: How nearly 9 million pounds of bad meat escaped into the food supply," narrates the story of how the toxic, "bad" meat leaked into the market. The mentality here resembles many other instances where a whole animal group/kingdom is killed due to one single infection, as was seen in the case of bird flu or hoof-and-mouth disease. The cause of disease is anthropogentic due to ill management, bad hygiene, and so on, and reflects human-centred bias and zoophobia. Such industrial farming practices, in addition to that of exotic animal consumption, are just a few examples of humanity's practice of scapegoating non-human animals for their own unethical treatment of the animal. Following this anthrosolipsist way of thinking, animals are inadvertently treated as the makers or accomplices of the apocalyptic world, rather than the primary victims of the global industrialisation of the meat industry.

This CNN report on animal toxicity elicits some thoughts about a contemporary, market-oriented mentality that perceives animals along the lines of production, consumption, and post-consumption. In further exploring the way economically value-less animals are victimised in global post-consumer society, in this section different ideas are investigated associating animals with waste in order to gain some insight into various forms of exploitation and suffering. Here, I follow Susan Signe Morrison in using waste as a broader umbrella category to encompass its sub-categories, such as trash and garbage (Morrison 2013: 464). I hope that, by providing some clarification of these waste-related synonyms in relation to non-human animals, specifically the words trash and garbage, it will become clear that different sets of animal–waste combinations or relationships evoke different epistemologies and affects for ensuing corresponding activism.

Let me begin with the term "trash animals" as a starting point. In *Trash Animals: How We Live with Nature's Filthy, Feral, Invasive, and Unwanted Species*, the authors define "trash animals" as the unwanted ones that are deemed worthless, destructive, threatening, and ugly, and they are despised as vermin, pests, nuisances, scavengers, and exotics. Trash animals are species-based and are often scapegoats for the naturecultural disasters in human communities. Each culture has its own trash animals; they include, to name but a few, coyotes, carp, pigeons, rats, and mosquitoes. These species, the authors point out, "are beginning to figure more prominently in the fields of animal-cultural studies and anthrozoology" (Kindle edition). In musing on the idea of "animal as waste," Randy Malamud defines this type of unwanted animal by equating it with "spoiled food, broken cell phones and computers, packaging, car carcasses—we can see that that, too, is being categorised, conveniently, as something we don't want around us, don't want to think about, don't want to deal with, don't want to integrate into a more challenging eco-vision" (Kindle edition). As much as we don't want to think about, to deal with, or to integrate with our waste, we treat the animals in exactly the same way as we see waste.

In other words, they are living trash because they are classified as such; their lives and species identity are labelled "trash" even before they "expire" and they need to be wiped out to safeguard the well-being of the human species. This "trash animal" can be viewed as an equivalent to the human outcast and Bauman's "wasted humans."

What, then, is the difference between "trash animals" and "garbage animals?" Are there any differences? And how would the difference help redress our deep-seated species bias and articulate a response? Before answering this, I shall clarify the difference between trash and garbage. First, the authors of *Trash Animals* state that "'Trash' commonly designates objects considered no longer useful or that have ceased to fulfill their intended purpose or benefit" (Kindle edition). They loosely define garbage as a variant of trash; in contrast to Morrison's classification, trash is in other words regarded as an umbrella term for other types of unwanted materials including garbage, refuse, filth, rubbish, and waste. However, the US government garbage officials define them differently. For example, in Madison, Alabama, their website defines trash as yard debris and large items, and garbage pertains to household waste.[6] Similarly, in *Rubbish! The Archaeology of Garbage*, the author defines trash as dry things such as paper and cans, and garbage as leftovers:

> *Trash* refers specifically to discards that are at least theoretically dry—newspapers, boxes, cans and so on. *Garbage* technically refer to "wet" discards—food remains, yard waste, and offal. *Refuse* is a more inclusive term for both the wet discards and the dry. *Rubbish* is even more inclusive: It refers to all refuse plus construction and demolition debris. (Rathje and Murphy 9)

Such a useful distinction between wet (leftovers) and dry (non-edible things) makes sense from a pre-industrial perspective because wet discards are used to feed animals, though it is no longer practised in the West. In Tuvalu, for example, garbage is divided into two types: edible and non-edible. The association of the word garbage with organic discards or food remains has a long history in the West.

Linguistically, the association of garbage with food, particularly offal, has to do with it first being used in referring to food waste in the fifteenth century from the Old French. According to the Oxford English Dictionary (2nd ed. 1989), garbage is defined as follows: (1) the offal of an animal used for food, esp. their entrails; (2) refuse in general; filth; and (3) fig. chiefly in the sense of worthless or foul literary matter (Scanlan 20;

Morrison 17). The Middle English word, garbage, refers to "giblets or parts of the fowls" which were served as a family supper during the early fifteenth century. In the sixteenth century, recipes containing offal started to lose their popularity, and its meaning began to become associated with filth, trash, or rubbish. From this discussion, we can see the word "garbage" is associated with animal parts and, therefore, justifies a further cross-examination between waste studies and critical animal studies.

While garbage, like trash and other "linguistic siblings" (i.e., rubbish and refuse, etc.), denotes a meaning of discarded materials, the technical usage of the word garbage as wet discards still preserves the etymological reference of animal corporeality and its evisceration: to disembowel the animals and take the viscera out of their bodies and either to savour them or toss them away as "garbage." Here, the animal's body is dissected into valuable/edible and worthless/inedible parts. In contradistinction to the label of "trash animals," in which living animals are conceptually compared to objects of modernity's excess (i.e., discarded cell phones), the label "garbage animal" activates an embodied epistemology. In a sense, it denotes that this group of animals undergoes an ongoing process of garbaging (first by turning into food, such as intestinal stew, and then by turning into filthy, unwanted discards). And in this process of becoming garbage, the gruelling parts of slaughtering and dismemberment are sufficient to arouse strong emotions and a unique interspecies aesthetic perception to potentially elicit passionate moral sentiments and, thus, trigger activism.

India's Sacred Cow as "Garbage Animal": *The Plastic Cow* and Documentary Activism

The Plastic Cow is part of "The Plastic Cow Project" initiated by the Karuna Society for Animals and Nature. After several mysterious animal deaths, they performed autopsies on them and discovered over 53 kilograms of trash and plastic products inside their bodies (Figs. 1 and 2). Clementien Pauws, the president of the Karuna Society, was convinced that these animals were the victims of plastic trash, and that the urban waste issue is an animal rights issue. "[T]he unholy alliance of plastic, carelessness and negligence," writes Philip Wollen, the president of Kindness Trust, "is not only torturing the Indian cow to death but it has also infected farming communities, rivers, forests and oceans, killing elephants, donkeys, fish, turtles and sea birds." Non-biodegradable waste, once swallowed, tends to stay in the stomach for decades and cause tremendous suffering and a

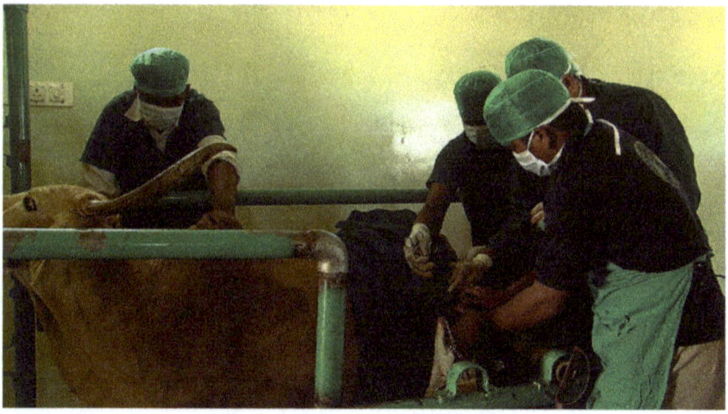

Fig. 1. A vet performing an operation to examine the contents of the rumen. A still from *The Plastic Cow*. *Source*: Courtesy of Kunal Vohra.

Fig. 2. After the operation, trash is collected and weighed. The total weight comes to 53 kilograms. A still from *The Plastic Cow*. *Source*: Courtesy of Kunal Vohra.

painful death. This project's ultimate goal is to ban plastic bags in India for the sake of the environment and the animals, and the film was meant to be one part of the campaign alongside a court case. Vohra writes, "In May 2012, soon after the film was made, a public interest litigation (PIL) was filed in the Supreme Court of India, as an animal rights issue. The case was admitted and the central and state governments across India, various ministries and municipal departments, etc., were made respondents" (email correspondence with the director).

In my view, while this documentary contributes to the activist, legal protection of the cows, as an art form it also provides a cinematic illustration of slow violence. The documentary's devices include narrative and aesthetic strategies, such as the employment of a meditative tone; a calm, philosophical, and reflective narrative style; long camera shots; and the incorporation of interviews. These elements help the audience to slowly absorb a form of slow violence. Furthermore, it presents this plastic cow issue not simply as that of animal rights but, more broadly, also as a universal problem of modernity and the global consumerist lifestyle in the context of Indian society.

The Plastic Cow is a sober, poignant, soul-searching journey into the heart of humanity, unveiling the glamour and hyperbole of Hindu religiosity to expose what lies beneath the veneer of lively rituals, tourism, and the bustle of city life: an invisible, slow, everyday cruelty against non-human animals, as well as against the land and environment. The voiceover states that India is caught between tradition and modernity, and cows have come to symbolise a paradox of sacred–profane hybridity at this particular historical juncture in Indian modernisation. At the beginning of the film, India is represented through the eyes of the tourist as an exotic sacred destination with its all too familiar theistic religiosity and spectacles (00:00–1:41). In a country awash with mythozoomorphic sensibility and imagination, we see an Indian woman gleefully proclaim the sacred status of the cows and their connection to Lord Krishna (3:38–3:55). In a culture with deep agricultural roots, where cows provide material/maternal sustenance, intense labour, and matriarchal religiosity, it is understandable that traditional Hindus consider cows to be their mothers. The ill-treatment (abandonment, illegal trade, slaughter) of cows can therefore be seen as a form of matricide. The bovine goddess, Kamadhenu, mother of all cows and iconographically depicted as an animal with a woman's face and a bovine body, inhabited by a host of different deities, is so sacred that it belongs to one of four symbols of Hinduism, and killing cows is a sacrilegious crime.[7] Reading the timeless matriarchal myth against the backdrop of contemporary Indian society, which has been invaded by aggressive, masculinist, transnational corporations such as McDonald's, which promote Western ideology and the practice of beef consumption, heralds the decline of bovine worship and respect. The continuation of the belief that cows are a symbol of wealth, abundance, and sacrifice in the context of global capitalism is shrouded in a darker, ironic connotation.

International and domestic consumption, and demand for beef and cow's milk, are the cause of the growth in the bovine population in India. In recent years, India has beaten Brazil to become the world's main beef (the Indian Buffalo) exporter, accounting for a fifth of global supply, as well as the largest producer of milk.[8] The domestic consumption of cow's milk is the primary cause of the widespread free-roaming of cows as they are "let go" by their owners after becoming unable to produce milk. Cows or buffalos, mostly aged or non-milking dairy cows, are banished to the streets and transformed into stray animals foraging amongst the garbage found in urban spaces. In particular, during the dry spring months, local farmers are unable to feed the cattle and "maintain cattle by letting them scavenge on the sparse grass along the roads." Marvin Harris writes, "In the cities cattle are encouraged to scavenge near food stalls to supplement their scant diet," and, according to Harris, these are the wandering cattle tourists report seeing throughout India (n.p.).

In the voiceover narration, the trope of paradox, commonly used to characterise modernity, is deployed to frame the narrative as well as to serve as a critical strategy to engage in deep thinking about slow violence and animal suffering. The philosophical tone, accompanied by the pensive Zen flute as background music, helps settle the mood to contemplate on this paradox in the age of global capitalism. Two Indian pundits offer answers: while one attributes the paradox to a loss of religious faith (4:20–4:38; 6:23–6:33), the other suggests a form of hypocrisy and ascribes it to the lies at the heart of religion (6:34–6:45).

Unlike in the case of traditional violence, in which shock value is at the core, such a moment tends to be absent in slow violence and therefore escapes media attention. However, this does not mean there is a lack of shock value-it just requires a more active pursuit as the effects tend to be hidden from view. As a matter of fact, the inception of The Plastic Cow Project and the subsequent documentary *The Plastic Cow* originates from the slow realisation of the kind of violence and cruelty that slips out of humanity's sight. After seeing case after case of animals, particularly cows, dying mysteriously, only to discover that the cause of their deaths was plastic waste, Pauws articulates her shock at the impact of "white pollution" (i.e., non-biodegradable plastics, etc.) on the animals (17:10–17:46; 20:28–21:23). And such a shock is vital as it registers on many critical levels.

The first level of shock is the simple and straightforward realisation that cows are dying as a result of munching on plastic bags. The second level of shock concerns the magnitude of the problem. In India, there are around 700 million free-roaming cows grazing on urban waste (it is not certain whether or not this figure includes temple cows, the cattle owned by urban farmers, and the cows that are let go after they are no longer productive). One can infer that almost all of the cows suffer from this slow, invisible violence. The third level of shock comes from the impact of the plastic cows on the production and consumption of the urban food chain. As mentioned above, there is strong demand for cow's milk and the domestic consumption of beef is also on the rise. The fourth level of shock is on a broader, global scale. *The Plastic Cow* can be added to the list of films highlighting the environmental impact of global mass consumption, like Chris Jordan's *Midway* project. Finally, on a meta-reflexive level, the shock is also one that demonstrates our ability to see suffering and violence in other species and, therefore, suspend what Žižek calls "fetishist disavowal" (45–46), that is, "I know, but I don't want to know that I know, so I don't know." Rather, it is, "Oh, my God, I hadn't realized this. Now what should I do to stop it?"

In "Testimonial Structures in Environmental Justice Films," Cory Shaman discusses Steve Soderbergh's *Erin Brockovich* (2000) and points out that Donna Jensen's moment of awakened consciousness "constitute[s] the political *point d'appui* for those who suffer from environmentally related disease" as it links all the issues of disease, toxic water, and corporate deception/abuse together. While in *Brockovich* the shift of consciousness is associated with "alternative health paradigms as environmental illness (EI) and multiple chemical sensitivity (MCS)," in *The Plastic Cow* it transforms an environmental waste issue into an animal rights issue. As previously mentioned, The Plastic Cow Project's ultimate goal is to ban plastic bags throughout India and, thus, to prevent animals from ingesting plastic products.

Another visual strategy that works to expose slow cruelty addresses the magnitude, complexity, and inter-relevance of events. Vohra ties the issues of modernity, animal rights, and urban waste management (or lack thereof) together. Moving on to the cause of the cows' deaths, i.e. the plastic junk they are exposed to, the director shows us the omnipresence of plastic bags on the streets and the magnitude of urban plastic waste production. Here, the documentary turns away from the idealised method of landscape representation for visual consumption. The mainstream tourist's impressions are represented by Hollywood's rhetorical conventions, which can frequently be found in the filming of Western landscapes by using such techniques as panning and close-up shots in order to render the land pristine. Beth Berila's discussion of Hollywood-style Western landscape and documentaries is useful in this context. In "Engaging the Land/Positioning the Spectator: Environmental Justice Documentaries and Robert Redford's *The Horse Whisperer* and *A River Runs Through It*," the author discusses the way in which panning (especially from above to create a colonial viewing position) and close-up shots are created to maintain the landscape as an object of sublime beauty, uninhabited by humans, rather than as "a living ecosystem of which humans are a part and on which we depend" (Berila 117). Berila alternatively evokes bell hooks' notion of "oppositional gaze" to read documentaries that "produce counter-narratives around environmental degradation and challenge viewers to read against the grain of dominant representations of nature in film" (117). So the choice of images and camera angles in the documentaries, at least in the case of *Toxic Racism* and *Drumbeat for Mother Earth*, are "much less celebratory of the landscape as a spectacle to be enjoyed from a distance; instead, they serve to implicate viewers in the issues portrayed in the films" (127). Similarly, in *The Plastic Cow*, the director challenges mainstream visual consumption of nature in film by producing a dark, slow spectacle: a view of the dumpsite slowly unfolds in front of the audience; instead of taking up a colonial viewing position that looks at the wasteland from a higher point, it uses eye-level panning and close-up shots. The slow, long panning shot of the wasteland lasts about 40 seconds. The use of both panning and close-up shots renders a mixed aesthetic experience (detachment and embeddedness) that echoes the theme of the puzzling paradox of modernity throughout the film. By juxtaposing the distant landscape of abjection alongside an interview with a local expert on the issue of plastic pollution and animal consumption of plastic bags, viewers' supposedly aesthetic or tourist gaze is turned into a close-shot view in which the audience is in the landscape, thus rendering a sense of immediacy. This strategy, of using both tourist and local shots, makes the issue at hand poignant to both local and global audiences. At the same time, it allows a cross-identification to take place (i.e., locals seeing their land from a panoramic tourist/outsider/ethnographic point of view and the global audience seeing the land from an insider's viewpoint).

Conclusion

This chapter has "ruminated" on the following themes: (1) false nature/culture or, more accurately, the civilisation/garbage divide; (2) different ideas of both humans and animals as worthless discards; (3) the linguistic connection between animals and garbage and the activist potential of the label "garbage animal," as well as the activist documentarian's use of moral shock; and (4) the fusion of environmental pollution and animal rights issues. In this context, waste/trash/garbage functions as an ecocritical site reflecting on its connection with disposable bodies, either human or non-human. The intersection between waste and animals contributes to current waste studies and also serves as a point of departure to launch a new integrated field of "animal–garbage" studies. An integrated, cross-pollination of compartmentalised fields helps broaden the horizon of current vision, which tends to be polarised towards environmentalism and animal studies. As in the case of *The Plastic Cow*, however, the issue sheds new light on critical animal studies, material ecological criticism, and transcorporeal studies.

A final note to consider is that the eco-documentary genre is a productive vehicle for engaging environmental and animal rights activism. In discussing the role that Indian documentary films play in the context of the environmental history of India, Rayson K. Alex admits that the medium of film has not been sufficiently considered as an effective tool to represent the environment and to bring about change in India. His assertion, which partially derives from the scarcity of environmental documentaries, as well as the limited number of venues available for screening environmentalist genres such as the documentary film, speaks for the importance of film festivals as vehicles of change. This, however, does not mean there has not been any success in bringing forth powerful political changes by virtue of film media. Taking Mike Pandley as an example, his *Shores of Silence* and *Vanishing Vultures* have helped to bring about significant legislative change. India banned the manufacture and sale of diclofenac (a veterinary medicine) in order to save vultures, thanks to *Vanishing Vultures*, which addressed this issue. Another documentary, *A Dance to Forget*, played a part in ending the practice of taming and training bears to dance for little roadside shows that were common in India for a very long time. These are some examples of documentaries that were "successful at making us care about large, global issues by making us care about the people involved, and by convincing us that if we are not ourselves activist and 'messengers,' we might still strive to be like them" (email correspondence with Vohra). As an unofficial side note, I was told that when *The Plastic Cow* was screened at TEFF, the reception was so overwhelming that the audience demanded a second screening! Similarly, when this documentary was shown to my students, the "shock value," particularly the scene of performing the autopsy, had a jaw-dropping effect when they saw about 50 kilograms of plastic retrieved from one cow's stomach. The "garbage animal," or the plastic cow, evokes such intense empathy that it prompts the audience to intuit the hidden animal suffering in the process of slow violence, which, in turn, can serve as a foundation of interspecies care ethics in the age of the Plasticocene.

Notes

1. See Morrison, *Excrement in the Late Middle Ages: Sacred Filth and Chaucer's Fecopoetics*. See also Sullivan, "Dirt Theory and Material Ecocriticism," and Serenella Iovino and Serpil Oppermann, *Material Ecocriticism*.
2. "E-Wasteland: Where do our electronics go at the end of their life": http://www.david-fedele.com/E-WASTELAND_files/ABOUT.html
3. Hance, "High-tech hell: New documentary brings Africa's E-waste Slum to Life." 30 April, 2012. http://news.mongabay.com/2012/0430-hance-ewasteland-interview.html
4. Description of *City's Edge* from the School of Media and Cultural Studies, Tata Institute of Social Science, Mumbai, India. http://smcs.tiss.edu/films/citys-edge/
5. In the case of AIDS, the main area of focus has been on the human dimension, whereas the root ecological problem of deforestation of the rainforest has not been given equal attention. See Anon, "Central Africa: Deforestation brings HIV/AIDS to indigenous communities, mainly women." *World Rainforest Movement*, July 2007.
   http://wrm.org.uy/oldsite/bulletin/120/Central_Africa.html
6. See the official website: https://www.madisonal.gov/FAQ.aspx?QID=82
7. For more information on the history of cow worship in India, see Marvin Harris' "India's Sacred Cow," http://sociology101.net/readings/Indias-sacred-cow.pdf
8. Weeks, Heather. "India: Milk's New Horizon": http://extension.psu.edu/animals/dairy/news/2013/india-milk2019s-new-horizon

References

Anon. "Central Africa: Deforestation brings HIV/AIDS to indigenous communities, mainly women." *World Rainforest Movement*, July 2007. Web. http://wrm.org.uy/oldsite/bulletin/120/Central_Africa.html

Barboza, Tony. "Researchers find plastic in more than 9% of fish in northern Pacific Ocean." *Los Angeles Times*. 2 July 2011. Web. 17 Aug. 2014. http://articles.latimes.com/2011/jul/02/local/la-me-garbage-patch-fish-20110701

Bauman, Zygmunt. *Wasted Lives: Modernity and Its Outcasts.* Malden, MA: Blackwell Publishing, 2004. Print.

Berila, Beth. "Engaging the Land/Positioning the Spectator: Environmental Justice Documentaries and Robert Redford's *The Horse Whisperer* and *A River Runs Through It.*" *Framing the World: Explorations in Ecocriticism and Film.* Ed. Paula Willoquet-Marcondi. Charlottesville, VA: U of Virginia, 2010. 116–34. Print.

City's Edge. Dir. Abhishek Yadav, et al. School of Media and Cultural Studies, Tata Institute of Social Sciences, Mumbai, India, 2012. YouTube.

Creed, Barbara. *The Monstrous-Feminine: Film, Feminism, Psychoanalysis.* New York, NY: Routledge, 1998. Print.

Fedele, David. Interview by Jeremy Hance. Web. 18 Aug. 2014. http://www.david-fedele.com/E-WASTELAND_files/ABOUT.html

Fox, Michael W. "India's Holy Cow: The Sacred and the Suffering." Web. 15 Aug. 2014. http://www.drfoxvet.com/library/special-reports/genetic-engineering-biotechnology-agriculture-and-food/india-sacred-suffering-holy-cow

Frates, Chris, and Shannon Travis. "Unfit for Human Consumption: How nearly 9 million pounds of bad meat escaped into the food supply." CNN, 2 May 2014. Web. http://eatocracy.cnn.com/2014/05/02/bad-meat-investigation/

Harris, Marvin. "India's Sacred Cow." Web. 17 Aug. 2014. http://sociology101.net/readings/Indias-sacred-cow.pdf

Iovino, Serenella, and Serpil Oppermann, eds. *Material Ecocriticism.* Bloomington, IN: Indiana UP, 2014. Print.

Johnson II, Philip David, and Nagy Kelsi, eds. *Trash Animals: How We Live with Nature's Filthy, Feral, Invasive, and Unwanted Species.* Minneapolis, MN: U of Minnesota, 2013. Print.

Koyaanisqatsi. Dir. Godfrey Reggio. IRE Productions, USA, 1982. Film.

Kristeva, Julia. *Powers of Horror: An Essay on Abjection.* New York, NY: Columbia UP, 1982. Print.

Madison, Alabama official Website: https://www.madisonal.gov/FAQ.aspx?QID=82

Manufactured Landscapes. Dir. Jennifer Baichawl. Foundry Films, Canada, 2007. Film.

Morrison, Susan Signe. *Excrement in the Late Middle Ages: Sacred Filth and Chaucer's Fecopoetics.* Basingstoke: Palgrave Macmillan, 2008. Print.

Morrison, Susan Signe. "Waste Aesthetics: Form as Restitution." *ISLE* 20.3 (Summer 2013): 464–78.

Murray, Robin L., and Joseph Heumann. "Introduction: Cinematic Eco-Disasters and Our Basic Human Needs." *Film and Everyday Eco-Disasters.* Ed. Robin L. Murray and Joseph Heumann. Lincoln, NB: U of Nebraska, 2014. Print.

Nixon, Rob. *Slow Violence and the Environmentalism of the Poor.* Cambridge, MA: Harvard UP, 2013. Print.

Plastic Bag. Dir. Ramin Bahrani. Noruz Film, USA, 2009. Film.

Rathje, William L., and Cullen Murphy. *Rubbish! The Archaeology of Garbage*. Tucson, AZ: U of Arizona, 2001. Print.

Scanlan, John. *On Garbage*. London: Reaktion Books, 2005. Kindle edition.

Shaman, Cory. "Testimonial Structures in Environmental Justice Films." *Framing the World: Explorations in Ecocriticism and Film*. Ed. Paula Willoquet-Marcondi. Charlottesville, VA: U of Virginia, 2010. 83–100. Print.

Song, Tian. "Global Warming as a Manifestation of Garbage." *Impasses of the Post-Global: Theory in the Era of Climate Change*. Vol. 2. Ed. Henry Sussman. Open Humanities Press. Web. 17 Aug. 2014. http://quod.lib.umich.edu/o/ohp/10803281.0001.001/1:6/--impasses-of-the-post-global-theory-in-the-era-of-climate?rgn=div1;view=fulltext

Sullivan, Heather I. "Dirt Theory and Material Ecocrititicism." *ISLE* 19.3 (Summer 2012): 515–31. Print.

Sullivan, Heather I., and Dana Phillips. "Material Ecocriticism: Dirt, Waste, Bodies, Food, and Other Matter." *ISLE* 19.3 (Summer 2012): 445–47. Print.

The Garbage Trap–Sanitary Workers of Indian Cities. Dir. Atul Pethe. Atul Pethe Productions, India, 2012. YouTube.

The Plastic Cow. Dir. Kunal Vohra. Altaire Films, 2012. Film.

Vohra, Kunal. Film director. Personal communication (email correspondence). 16–26 June 2014.

Weeks, Heather. "India: Milk's New Horizon." *Penn State Extension*. Web. 17 June 2013. http://extension.psu.edu/animals/dairy/news/2013/india-milk2019s-new-horizon

Žižek, Slovaj. *Violence: Six Sideways Reflections*. London: Picador, 2008. Print.

Surviving Progress, Modernity and Making Sense of the Crisis in Nature

Divya Anand

Progress. Standing at the crossroads of meaning in the twenty-first century, the term is increasingly contested with regard to its positive etymological connotations. The historical understanding of progress as economic, social, and cultural development is undeniably tied to the way in which humankind has engaged with nature throughout time. And any analysis of nature should begin with the understanding that with every human encounter with nature, we socialise with nature and make it meaningful. Today, there is hardly any physical space on earth that is untouched or unknown to human beings. While nature is perceived through social mediation—both discursive and real—it also remains true that nature can exist outside of human beings' perception and awareness.

The looming and perceptible catastrophes of climate change and global warming point towards the dissonance in the ways in which nature is socialised. The future of humankind is increasingly foretold as the apocalyptic "end of nature" (McKibben, *The End of Nature*) or "after-nature" (Strathern, *After Nature*). If human beings continue with the current status quo of engaging with nature, then that very nature will cease to sustain humankind. This is not the "end of nature" but the end of nature "as we know it" (Flannery, *The Future Eaters*). The idea of "risk" to nature

D. Anand (✉)
Diversity and Academic Programs Associate, Novartis Institutes for BioMedical Research, 932 E, 200 Technology Square, Cambridge, MA 02139, USA

R.K. Alex, S.S. Deborah (eds.), *Ecodocumentaries*,
DOI 10.1057/978-1-137-56224-1_7

(Beck, *Risk Society*), in this sense, is the risk of our survival in nature, and this risk is understood and constituted by social, political, and historical factors acting together. The spectrum of responses, ranging from pressing concern to indifference towards nature, is also rooted in the different ways in which nature is engaged with, and understood from, contrasting social, political, and geographical standpoints. The radical increase in the ways in which nature is transformed and, more often, irrevocably so, also means that the effects of such transformation of nature are borne unequally by people and societies across the world.

Globalisation has meant increasingly interconnected networks between natural resource extraction and the consumption of finished goods, which exacerbates the ecological footprints of humankind. At the same time, alleviating the exponentially increasing impacts of the consumerist–capitalist societies requires collective responses springing from a shared concern for the environment. Such a concern for the environment has to overlap arbitrarily drawn borders and boundaries of physical space, most definitively those of the nation-state, as ecosystems overlie such borders and solutions to ecological crises can only be addressed through bilateral collaborations. The continuing damage and struggle to find and implement appropriate solutions to protect and preserve nature "as we know it" invariably leads to a tension that is palpable in the contesting claims to nature, from small communities to the nation-state to international corporations as stakeholders.

The Enculturation of Nature

The processes of enculturating nature, which started with hunting and gathering, have progressed to the new sciences of genetic engineering. For example, Tim Flannery in *Future Eaters* (1994) has recorded how environmental changes over millennia occurred through the complex interaction of variant factors on the Australian continent. This led to upsetting the balance of the ecosystem and the extinction of many mega-herbivores during Aboriginal inhabitation of the continent, leading to the adaptation of these changes into a new ecosystem by the time of European settlement. Alfred Crosby's account of ecological imperialism, starting from the end of the Ice Age, when humans began exploring the Eastern Atlantic, to the creation of "neo-Europes" in North and South America, Australia, and New Zealand, attests to the alteration of ecological and biological landscapes (*Ecological Imperialism* 7). The recent discovery of plastic-eating bacteria also points to the organic process of evolution and adaptability to changes in the environment. Be that as it may, the "crisis of nature" is tied to the accelerated processes of enculturation associated with industrialisation and its aftermath. Explorations and the discovery of the "new worlds," followed by colonial expansion, were, to a great extent, coeval with industrialisation. Subsequent political changes and increased traffic and communication around the world meant radical changes in the ways in which nature was socialised.

It was Theodor Adorno and Max Horkheimer (the Frankfurt School) in *The Dialectic of Enlightenment* (1947) who pointed out the flawed logic in the human mastery of nature, which they identified with the totalitarian regimes of the twentieth century. Adorno and Horkheimer located the crisis of modernity in the rise of the state's power, combined with the increasing domination of, and disassociation from, nature through science and technology. The emancipatory potential that Karl Marx had envisaged by improving the means of production through the subordination of nature was, for Adorno and Horkheimer, the very reason for the human subordination they witnessed in their native Germany and in modern Europe. Increasing industrialisation meant the objectification of nature and a growing objectification of human beings as part of nature itself. This chasm of the increasing objectification of nature and "the liquidation of the individual" was the difference between pre-modern and modern societies, which they feared led to exploitative relationships of dominance and subordination among human beings. Nature and those "closer to nature" were then easily classified as the "Other" to be dominated. In Adorno and Horkheimer's indictment of the totalitarian effects of the dialectic of Enlightenment in modernity, David Roberts and Peter Murphy discern an appeal to romantic reconciliations of nature and society or the "dialectic of romanticism." According to Roberts and Murphy:

> Enlightenment autonomy is always threatened by the immanent contradiction of denaturalization: the reversal of freedom into unfreedom (the perpetuation of the blindness of nature); romantic incarnation is always threatened by the immanent contradiction of renaturalization: the reversal of the spiritualization of nature into the naturalization of spirit, of creative into destructive nature. Each bears witness to the failed internal dialogue of modernity. (*Dialectic of Romanticism* 75–76)

Roberts and Murphy argue that the modern impulse to resurrect the "immanence of nature" was "directed not only against the modern project of the unlimited growth of knowledge, production and power, but also against the universal project of democracy" (*Dialectic of Romanticism* 68). The logic of immanence in nature, which "reduces reason to life and history to the biological cycle of the growth and decay of cultures, becomes a deadly logic of modernity when particularism becomes nationalism and race is raised to an exclusive and absolute integrating value" (*Dialectic of Romanticism* 70). So both the dialectic of Enlightenment and the dialectic of romanticism were tied to modernity in ways which undermined their professed objectives of human freedom and equality. Agnes Heller has also noted that the human stance that treats nature as a "standing reserve" is as much a product of modernity as the adoration of nature for its own sake. She adds that "one might say that the more nature has come to be regarded as a mere object, a 'standing reserve' for human use, the more it has become beautiful to 'the eye of the beholder', who (disinterestedly!) lets nature stand where it stands" ("Three Logics of Modernity" 69–70). The dialectic of romanticism remains manifest in "aesthetic nationalisms" or "national mythologies" inhering with the idea of the "imaginary" in promising an organic totality, whether by entreaties to blood and soil or common ties of language, history, or race. The contradictory impulses of Enlightenment (belief in reason and science) and romanticism (idea of roots and soil) were both utilised in legitimating the nation. Correspondingly, the idea of the imaginary also explains how nations and nationalisms are repositories of contradictory elements, which can be rallied for both reactionary and revolutionary, homogenising and heterogenising, and assimilatory and discriminatory tendencies. If the enculturation of nature, identified as "risk," is a consequence associated with the social forms of modernity—industrialisation, capitalism, science—solutions to address this "risk" are also products of modernity.

Defined as "development that meets the needs of the present without compromising the ability of future generations to meet their needs" (WCED 1987: 43), the concept of sustainable development was posited as a solution to the excesses of industrialisation and the consequent exploitation of nature at unsustainable levels. The term emerged as an elastic concept with multiple interpretations and varied practices, most of which have undermined the sustainability aspect for the sake of development. The appropriation of "sustainable development" by corporate interests has led to the redefinition of the term, even as new concepts and practices have emerged to address the reckless pace of environmental degradation in the twenty-first century. De-development, post-development (Bennett 2012; Ahorro 2008), de-growth (Martinez-Alier et al. 2010; Latouche 2009), a-growth (Van den Bergh 2011), post-growth, the slow movement (post-growth.org, Honore 2004), and radical social democracy (Shrivastava and Kothari 2012) are but some of the varied conceptualisations that have emerged from the recognised failure of the spirit and practices of the post-war development ethos. These concepts are bound by the urge to re-imagine nature in ways that slow down natural resource extraction and critique the unidimensional nature of post-war development, which continues to be the driving mantra of industrialising and post-industrial societies.

The radical development in science and technology, which ushered in industrialisation, colonisation, nation-building, globalisation, and, at the far end of the scale, global warming and climate change, is a part and product of modernity. In the trajectory of the evolution of the modern technological world lies an inherent paradox that the same factors that are causing the impending catastrophes of nature can also cure them. This paradox is intrinsic to modernity, in the sense that while industrialised modernity is the cause of environmental damage and rising social inequalities, it can, at the same time, eradicate social inequalities and environmental damage through its promise of human freedom.

Progress and Its Discontents

The idea of progress is intertwined with the paths and goals of modernity, where developed societies occupy the pinnacle of progress which is coveted and pursued by the rest of the world. Progress, in this sense, is best understood as part of the "social imaginary." Cornelius Castoriadis in *The Imaginary Institution of Society* is one of the first scholars to identify an imagined aspect to every society, which exists through "the whole complex of its particular institutions" (6). Each society collectively and creatively constitutes its own networks and institutions, which sets it apart from other societies. Charles Taylor writes of social imaginaries as "the ways people imagine their social existence, how they fit together with others, how things go on between them and their fellows, the expectations that are normally met, and the deeper normative notions and images that underlie these expectations" (*Modern Social Imaginaries* 23). The idea of progress, that each society aspires to and distinguishes itself from, exists in the realm of the social imaginary.

Harold Crooks and Mathieu Roy's documentary, titled *Surviving Progress*, focuses on the many different narratives of progress from beginning to end. Inspired by Ronald Wright's *A Short History of Progress*, the documentary begins by defining the term progress, and leads to the causes and outcomes of a particular idea of progress. This idea has won universal traction and remains the foundation of many of the world's social and environmental problems. Progress is also a word upon which ideas of nationhood and identity have been built and buttressed, wars fought and forests plundered, rivers dammed, people dispossessed and displaced, and plants and animals driven to extinction (Diamond, *Guns, Germs*). For example, approximately 45,000 dams were built the world over following the Second World War, where the gargantuan structures controlling and redirecting water became markers of national pride. The Narmada Valley Development Project in India, the Tennessee Valley Authority in the USA, and the development of the Mekong River in China are but a few examples of structures showcasing the scientific and technological expertise of nations—the badges of pride and progress erected for the greater common good. The continuing impulse to build dams in the developing world, at a time when dams are increasingly decommissioned and river waters restored in the developed world, is testimony to the continued traction of such yardsticks of progress (Worster, *Rivers of Empire*; McCully, *Silenced Rivers*). Progress is thus the basis upon which an individual's and a nation's success is measured. And for both people and nations, it is in the accumulation and enjoyment of material wealth that progress is accomplished.

Surviving Progress examines the idea of progress, from the physiology of humankind from 50,000 years ago to the twenty-first-century human genome. Interpolated with commentary by authors such as Ronald Wright and Robert Wright, economic historian Michael Hudson, primatologist Jane Goodall, writer Margaret Atwood, former Brazilian environment minister Marina Silva, biologist Craig J. Ventner, and activists such as David Suzuki and Kambale Musavuli, *Surviving Progress* cuts an arc across the common understanding of, and the paradoxical quality of, progress, as manifest in different parts of the world. The melange of images and commentary is ingeniously drawn together to provide a nuanced and in-depth insight into the differing and often contradictory aspects of our paths, goals, and understanding of the current crises of nature. *Surviving Progress* can be seen as representative of the many documentaries exploring the reasons behind environmental degradation and gravely cautioning humankind of its consequences. In its narration from multiple vantage points, as it weaves in many timelines, *Surviving Progress* offers a rich critique of the current relationship of humans with nature, opening up the question as to what is the best way to communicate with humanity and to sensitise us to rethink our path forward into the future.

Progress and Its Traps

The documentary is premised on what Ronald Wright calls the "progress traps"—human behaviours which initially are productive but progressively culminate in disasters. The example of the mammoth hunters, who discovered that driving the herd over a cliff would yield immediate and immense results which, however, backfired when this practice led to the extinction of mammoths, typifies a progress trap (01:07:32). The urge of humankind to reap benefits at an exponential rate, without thinking about the long-term implications of their actions, is the main thrust of this idea. In the modern world, it is the unbridled faith in technology and science that has culminated in the current world crisis, which is the "progress trap." The similarity between the Occupy Movement (a social movement against corporate greed which began on 17 September 2011 in New York and which has spread to over 1500 cities globally) and *Surviving Progress*, as Mike Scott rightly points out, is that it "intellectualizes much of that movement's anti-greed message and paints a portrait of a civilization with an end-game eerily similar to that of the Roman empire" ("'Surviving Progress' review").

As Ronald Wright makes clear, one of the major paradoxes of the crises of humankind stems from the fact that our physical bodies and our physical brains remain more or less the same as they were in the Stone Age, but our culture has evolved exponentially, moving far ahead of our natural pace of evolution. So while humans have the capacity and the wherewithal to create and decimate, we lag far behind in terms of our cognitive abilities to discern and decide at such levels of sophistication and complexity. On the other hand, it is our capacity to question, "to ask why," as David Povinelli voices in *Surviving Progress*, which sets us apart as intellectual beings with the ability to invent modern technology (01:15:20).

The key word here is modern. While humans have evolved for the past 500,000 years, it is only in the past 200 years that "progress," in the contemporary sense of the word, starts to make sense. Progress, which is ushered in by the advent of modernity. Progress through the pursuit of science and technology was equated with greater human freedom and equal opportunity. It coalesced with the idea of nature as an endless resource to be utilised for the production of wealth and the progress of human beings, along with according primacy to reason and scientific knowledge. As Anthony Giddens remarks, the negative characteristics of modernity were viewed as outweighing the benefits proffered by modern social forms, by sociological thinkers from Karl Marx and Emile Durkheim to Max Weber. Two negative consequences of modernity that Giddens points to are the large-scale destructive potential of the "forces of production" upon the material environment and the materialisation of totalitarian forms of government (1991: 7–8). For Karl Marx, the "means of production"—the remaking of nature with science and technology—was the key factor that would help remake society into a just and equal one. Marx conceived humanity's relationship with nature, historically developed through capitalism, as having an emancipatory potential that could only be realised through a transformation to socialism and then to communism (McLaughlin 1993: 118; Schmidt 1971; Castree 1995; Morrison 1994: 133; Mingione 1993: 86; Benton 1989: 76). According to Marx, improved and efficient means of production—through industrialism—would reduce the need for human labour, leading to more free time to build an equal and free society. It is in the Marxist view of nature as a resource to be extracted for human well-being through industrialisation that the common denominators of socialism and capitalism are identified. While Marx, following the Enlightenment ideals, placed his faith in reason, science, technology, and the human mastery of nature, equating it with human progress, it was not an uncritical suspension of judgement towards science, technology, and ideas of progress (Biro, *Denaturalizing* 89). Marx's conviction that technology can be deployed in the service of human beings to enhance the means of production inadvertently meant that the domination of nature raised the possibility of domination among human beings themselves. Partha Chatterjee's delineation of the connections between the sciences of nature and colonisation in the nineteenth century makes this correlation apparent. Chatterjee avers that in post-Enlightenment Europe, the sciences of nature became the paradigm of all rational knowledge, based on the idea that the sciences of nature facilitated man's control over nature to serve human interests. Serving human interests revolved around the concepts of wealth, progress, efficiency, and so on, which were defined in terms of "promotion of social interests." Since "interests" in societies vary and are apparent in terms of relations of power, "the subject-object relation between man and nature is now subtly transferred, through the 'rational' conception of society, to relations between man and man" (*Nationalist Thought* 14). In this manner, sciences of nature become sciences of society, which in turn become the "knowledge of the Self and the Other [and] in terms of rationality, it necessarily also becomes a means to the power of the Self over the Other. In short, knowledge becomes the means to the domination of the world" (*Nationalist Thought* 14–5). The aspiration of colonial nationalism struggling against the empire was also derivative of scientific and rational knowledge, drawn from the empire and the logics of modernity.

The pursuit of progress by more than half the earth's people in the developing world is imagined and engineered towards a model of development that is western and "developed." The film shots of Chen Ming, the Chinese driving-club tour guide, forbidding his father from citing the environmental costs of China's development, encapsulates the dilemma of the aspirations of the majority of humanity to pursue progress while turning a blind eye to its environmental costs (02:05:34). As Ulrich Beck writes in "Climate for Change," nations, in treading the path of "becoming more westernized, reflect the West back to itself, so that the 'equality' of environmental destruction leads to the self-destruction of civilization" (257). The alternating shots of commentary by Ronald Wright and interdisciplinary scientist Vaclav Smil, and the visuals of Ching Men and the ex-Chinese government interpreter placing industrialisation and development following the West as China's turn to the "right path" explicitly show the convergence of two radically different nation-state ideologies in the name of progress.

Examining the idea of progress and development vis-à-vis emerging nation-states in the twentieth century also reveals two different paths of modernisation, signified by liberal–democratic–capitalist societies, such as the USA, and communist societies with autocratic governments, such as China. Following on from the industrial revolution, liberalism became the chief impetus in the West through the market economy, and socialism in the East by a planned economy, with both sharing the same goals of growth and development. Cornelius Castoriadis locates two competing imaginaries in liberal democratic societies, that of the market and the autonomous society. The eastern models of society constituted a second version of modernity characterised by the strato-bureaucratic imaginary, where the state and the bureaucracy were vested with powers to "develop" the nation. The "communist project of modernity" was a distinctive version of modernity, or "alternative modernities," seen in the evolution of a "Soviet model" of development. Both capitalist and communist societies shared a common impetus towards growth and development, which marks modernity, although the institutions of society and their imaginaries functioned differently in the pursuit of this ideology (Castoriadis, "From Ecology to Autonomy" 41). This meant that the dominant social imaginary in both socialist and capitalist countries was founded on the same tenets of growth and development, where nature was to be mastered through scientific and technological innovations.

China's aspirations for progress, following those of the USA, also showcase the quandary of balancing human ideas of progress with the finitude of natural resources. The images of the New York Stock Exchange and clips of the speeches by Ronald Reagan and George W. Bush during their presidencies, the former promising to let the "bull run lose" in terms of freeing the market economy from governmental regulation and the latter, articulating optimism in the same free markets following the 2008 economic recession, underscores this point. Inserted between the shots of Reagan and Bush is the Chinese speaker endorsing the same idea of economic growth as the cornerstone of progress. The transformation of China into a "socialist market economy," accelerated since the 1990s, followed a policy of uneven development creating and intensifying regional inequalities (Wei, *Regional Development* 3). The dramatic rise in the quality of life of people, such as Cheng Min featured in the documentary, and the tourists he guides, with access to seasonal fruits throughout the year, contrasts with the poverty of the villages he tours, where no one owns a refrigerator and one has to walk 30 kilometres to buy meat once a week. China's march towards a market economy, while yielding a rapid rise in its gross domestic product (GDP), has also dramatically exacerbated social and economic inequalities (Shi et al., *Rising Inequality*). If China shows both sides of the story of the free run of the market, the global effects of letting the "bull run loose" is starkly tied up with the snapshots of Brazil and Congo.

In Brazil, the election of neoliberal politicians and the massive flow of foreign capital as loans, from the 1990s onwards, facilitated by the International Monetary Fund (IMF) and the World Bank, set the stage for the forfeiture of its vast natural capital to foreign stakeholders. Neoliberal policies of developing nations setting their sights on progress meant that foreign capital was solicited and handed over, with interest rates set by global banking institutions. As James Petras notes, it is the election of the neoliberal President Fernando Henrique Carduso in the 1990s that set Brazil on the path of accelerated privatisation and deregulation (1), a decade when many developing nations, such as China and India, treaded the same path. The Amazon forests sustain 13 % of the world's biota and also have one of the highest deforestation rates in the world, averaging 0.52 % annually (Sarber, "Roads, Deforestation" 203), which remains high despite government policies finding a measure of success in the past decade. And for the Brazilian Federal Police spokesperson in *Surviving Progress*, monitoring logging practices also underlines the issue that it is not the locals cutting down trees for their livelihood which poses the major threat to the Amazon, but the actions of big agro-business owners and cattle ranchers, the majority among them being Brazilian politicians in power. The shot of a clear-cut area of the Amazon, sans background music, drives home the point further. The school child espousing the simple logic that "If we don't take care of the Amazon forest, it will become the Amazon desert because when one cuts trees in a forest one creates a desert" articulates the simple equation of deforestation and its direct effects (*Surviving Progress* 03:16:13). The corporatisation of natural resources where ownership, production, and consumption are transnationalised also helps circumvent accountability in the wake of the huge social, economic, and ecological costs.

The trinity of the IMF, the World Bank, and the World Trade Organization (WTO) has, since the 1990s, held the torch of economic globalisation, wherein these organisations have railroaded loans and facilitated the access of multinational corporations to the natural resources and cheap labour of low-income countries, while holding their national economies at ransom for excessive interest rates. Walden Bello and Anuradha Mittal have termed the transnational banking organisations as "unmitigated disasters" with their oligarchic decision-making processes, centralising tendencies, and granting free access and controlling power to big corporations (*The Future in the Balance* 2001). *Surviving Progress* starkly captures with great poignancy the unfairness, inequality, and unbalance between the two stakeholders, with the black-and-white clip of a suited businessman in a posh high-rise office room cautioning the person he speaks to with the words:

> Look, don't give me a hard luck story, I hear them every day and quite frankly, they bore me. The facts are simple: In 1973, this bank gave you a loan, and you still haven't paid it back. Admittedly, you paid back the initial sum, but not the interest, which, to date, amounts to 9 times the amount originally borrowed. Nine times. So you better get your act together. Times are tough, and we're all having to clamp down. And don't look at me like that. This is a bank, not a charity. (*Surviving Progress* 03:03:04)

The camera zooms in to show a hapless African boy sitting at the other end of the table. The directness of this juxtaposition poignantly captures the inequality and the unjust nature of financial and corporate transactions that have happened since the 1980s, its devious perpetrators, and its innocent victims.

The Democratic Republic of Congo (DRC) showcases another angle to the murky mix of nationalism, globalisation, and corporatisation, which followed the rule of Mobutu Sese Seko from 1965 to 1997. Aided and abetted by the USA and its former coloniser, Belgium, Mobuto ran a kleptocracy where his authoritarian rule, propped up with enormous foreign funds, inadvertently kept its vast forests and people in their "undeveloped" state. While this meant the rich forests remained relatively unscathed, the people of the DRC suffered in terms of a lack of basic facilities in education and healthcare, corruption, and large-scale human rights violations. The subsequent civil war and political unrest meant that the country continued in its wretched state and its rich forests remained protected. However, retrieving the international debts incurred by previous governments, and political stability from the 2000s, meant the incursion of international stakeholders, including the IMF and the World Bank, for both debt reclamation and "development." The extent and rate of deforestation in the Congo has since accelerated, while human rights violations continue unabated (Ernst et al., "National Forests"). Ranked among the lowest of countries in terms of Human Development Indices, the DRC powerfully symbolises, in multiple ways, the long-term effects of the ideology of progress steered by multinational interests.

The almost universal pursuit of this particular kind of development and progress generates inequalities at a multi-scalar level, both within nation-states and between nation-states. The globalised free market increases both the expanse of production and the number of consumers, in turn intensifying social and economic gaps between people, accompanied by unmitigated ecological damage. The question raised here is how and why do ideas of progress have such universal appeal, wherein today, despite knowing the heavy ecological and social costs, people and nations tread the same paths of progress.

The idea of an oligarchy was traditionally attached to failed or corrupt governments in the developing world, whereas today, given the disappearance of the middle class and the rising number of poor, the term oligarchy can equally be applied to the USA. In a study based on surveys taken from 1981 to 2002, Martin Gilens and Benjamin I. Page conclude that the USA, while granting democratic rights to its citizens, such as freedom of speech and the right to vote, functions more as an oligarchy wherein economic elites and organised business groups have substantial influence over government policy in comparison with the average citizen and public interest groups (Gilens and Page, "Testing Theories" 594). The economic recession of 2008 and the remarkable recovery of Wall Street reiterate this point. In *Surviving Progress*, the former economist of the IMF, Simon Johnson, describes Wall Street's hold over the US government, both pre- and post-2008 recession. Also, while globalisation has created networks of production and consumption across the world, it has also generated an increasing number of the global poor, who transcend the divisions between the developed and the developing world. To start with, we need to identify the nature and scope of the problem which Marina Silva, the former Brazilian Minister of the Environment, articulates in *Surviving Progress*:

> I think we have reached the era of limits. Although we are free, we must live within the limitations of nature. It is impossible to defend models that cannot be universally applied because we would have to start from a premise that some people have rights and some don't. Thus there is no technological problem, but an ethical one. (03:09:26)

The next question in the viewer's mind is "How do we solve this problem?" and it is in answering this question that *Surviving Progress*, on many levels, flounders.

Conclusion

While *Surviving Progress* fuses together variant strands of thoughts and philosophies in order to drive home its point, the richness of this tapestry is, at times, overwhelming for the viewer. The diverse range of scholars and disciplines coming together to frame this grand narrative would have benefitted from a stronger presence of dissenting voices. With its over-abundance of questions spanning numerous subjects, disciplines, timelines, and geographies, it strikes an extremely pessimistic note that falls short of being counterbalanced in terms of the way forward. The paradox here is that while industrialised modernity is the cause of environmental damage and rising social inequalities, at the same time, it holds the key to the eradication of social inequalities and environmental damage with its promise of human freedom. So film clips of a man on the moon, megalopolises, robots, and gargantuan chimneys emitting noxious plumes, on the one hand, appear as dystopian but, on the other, they are also images of a utopia, propped up by our ideas of development and progress.

What saves *Surviving Progress* from becoming a dark, doomsday, prophesying narrative is the fragile optimism in offering a solution to the current crisis of the earth—consume less. In critiquing the pace at which developing nations like India, China, and Brazil are racing towards the levels of consumption of the developed world, placing the onus of "consuming less" on such countries also hides a moral double standard wherein developed nations simply carry on consuming and serve no reparations for the prosperity they have enjoyed in the past, a prosperity which was founded on the twin fulcrums of industrialisation and colonisation. This uneasy moral quandary briefly comes to the fore when the Chinese Tour Operator admonishes his father for highlighting the drawbacks of his, and in turn, China's, new-found prosperity (02:05:17). The documentary, however, does not dwell further on this subject nor offer any new solutions or insights into the way forward for those parts of the world where the environment is struggling. Whether we call it progress or not, the way forward is an issue that remains largely unaddressed. As Lee Parsons notes:

> On balance, we are invited to take part in a collective self-condemnation and even despair, while any deeper understanding of the historic crisis humanity now faces is said to be insoluble on the basis of further social development. Whatever the intention, this leaves quite specific social groups, i.e., the financial and corporate elite, off the hook. If the problem is *everyone's* fault, then it is effectively *no one's*. (2012)

Furthermore, how is this documentary on the environment different from the many others that are churned out every year? How does it set itself apart from the numerous doomsday narratives characteristic of environmental documentaries from the recent past? The answer is, it doesn't. While *Surviving Progress* stands apart in seamlessly combining staggering visuals with compelling narration, echoed perfectly by the background score in its sweeping survey of "progress traps," it fails to go beyond the very dramatised "why," that is, why and how have we reached this point of ecological damage? To survive is to consume less, but how? That is a question which largely remains unexplored in *Surviving Progress*.

The current equation of profit with progress, acquired by the 1 % at the helm of corporations, also glosses over the historical and social underpinnings of the market economy. *Surviving Progress*, while spanning the length and breadth of the development of the human race, de-historicises the "market" by overlooking the specific rise and role of the market economy in particular points of time, especially the role of the market in dislodging feudal societies. Also, the broad generalisation of the current crisis negates the many small-scale and micro-sites all over the world where redemptive measures have successfully salvaged and addressed social and ecological problems in innovative and pioneering ways.

Surviving Progress revisits with powerful imagery and with words from the world's eminent experts from divergent streams. In weaving this grand tapestry and bringing together the various disciplines of science, history, geography, politics, and biology lies the strength and beauty of the documentary. It is caustic and grandiose in its verbosity and dramatisation, a contingency that is called for in terms of the seriousness of the issues discussed. It is also a potpourri of ideas, thinking minds, music, and imagery that is optimally juxtaposed and blended in, which effectively entertains and drives home the point—we live in a finite world with no alternative in sight and we must dramatically rethink and rectify our current ways of living if we want a future on the earth.

References

Ahorro, J. "The waves of post-development theory and a consideration of the Philippines." Edmonton: University of Alberta, 2008. Web. www.cpsaacsp.ca/papers-2008/Ahorro.pdf

Barber, C.P., et al. "Roads, deforestation, and the mitigating effect of protected areas in the Amazon." *Biological Conservation* 177 (2014): 203–09. Web.

Beck, U. *Risk Society: Towards a New Modernity*. New Delhi: Sage, 1992. Print.

Beck, U. "Climate for Change, or How to Create a Green Modernity?" *Theory, Culture & Society* 27.2–3 (2010): 254–66. Print.

Bello, W.F. *The future in the balance: Essays on globalization and resistance*. Ed. Anuradha Mittal. San Francisco: Food First, 2001. Print.

Bennett, C. "Supporting the Posts in Development Discourse: Under-development, Over-development, Post-development." *Sociology Compass* 6. 12 (2012): 974-986. Print.

Benton, T. "Marxism and Natural Limits: An Ecological Critique and Reconstruction." *New Left Review* 178 (1989): 51–85. Print.

Biro, *Denaturalizing Ecological Politics: Alienation from Nature from Rousseau to the Frankfurt School and Beyond*. Toronto, Canada; Buffalo, NY; London: U of Toronto, 2005. Print.

Castoriadis, C. "From Ecology to Autonomy." *Thesis Eleven* 3 (1981): 8–16. Print.

Castoriadis, C. *The Imaginary Institution of Society*. Cambridge, MA; London: MIT Press, 1987. Print.

Castoriadis, C. *World in Fragments: Writings on Politics, Society, Psychoanalysis, and the Imagination*. Ed. and trans. D.A. Curtis. Stanford, CA: Stanford UP, 1997. Print.

Castree, N. "The Nature of Produced Nature: Materiality and Knowledge. Construction in Marxism." *Antipode* 27.1 (1995): 12–48. Print.

Chatterjee, P. *Nationalist Thought and the Colonial World: A Derivative Discourse*. London: Zed, 1986. Print.

Ernst, C. et al. "National forest cover change in Congo Basin: Deforestation, reforestation, degradation and regeneration for the years 1990, 2000 and 2005." *Global Change Biology* 19.4 (2013): 1173–187. Print.

Giddens, A. *The Consequences of Modernity*. Stanford, CA: Stanford UP, 1991. Print.

Gilens, M. and B.I. Page. "Testing theories of American politics: Elites, interest groups, and average citizens." *Perspectives on Politics* 12.3 (2014): 564–81. Print.

Heller, A. *A Theory of Modernity*. Malden, MA: Blackwell, 1999. Print.

Heller, A. "The Three Logics of Modernity and the Double Bind of the Modern Imagination." *Thesis Eleven* 83 (2005): 63–79. Print.

Honoré, C. In praise of Slowness: How a Worldwide Movement is Challenging the Cult of Speed. New York: HarperCollins, 2004. Print.

Martínez-Alier, J. et al. "Sustainable de-growth: Mapping the context, criticisms and future prospects of an emergent paradigm." *Ecological Economics* 69.9 (2010): 1741–1747. Print.

Kaika, M. *City of Flows: Modernity, Nature and the City*. London: Routledge, 2005. Print.

Latouche, Serge. *Farewell to Growth*. Polity, 2009. Print.

McKibben, B. *The End of Nature*. London: Viking, 1990. Print.

McLaughlin, A. *Regarding Nature: Industrialism and Deep Ecology*. Albany, NY: State University of New York, 1993. Print.

Mingione, E. "Marxism, Ecology and Political Movements." *Capitalism, Nature, Socialism* 4.2 (1993): 85–92. Print.

Morrison, R. "Two Question for Theory and Practice: Can you be Marxist and Green?" *Rethinking Marxism* 7.3 (1994):128–36. Print.

Parsons, L. "Surviving Progress: A dim view of humanity." Review of *Surviving Progress*. World Socialist Website. 3 Feb. 2012. Web. https://www.wsws.org/en/articles/2012/02/prog-f03.html

Petras, J. "Brazil: Extractive Capitalism and the Great Leap Backward." Global Research. 23 July 2013. Web. http://www.globalresearch.ca/brazil-extractive-capitalism-and-the-great-leap-backward/5343624

Sato, H, K. Kuribayashi, and K. Fujii. "Possible practical utility of an enzyme cocktail produced by sludge-degrading microbes for methane and hydrogen production from digested sludge." *New Biotechnology* 33.1 (2016): 1–6. Print.

Schmidt, A. *The Concept of Nature in Marx.* Tr. Ben Fowkes. London: New Left Books, 1971. Print.

Scott, M. "'Surviving Progress' review: Documentary sounds an alarm for humanity." 24 May 2012. Web. NOLA.com-The Times-Picayune

Shi, L. et al., eds. *Rising inequality in China: Challenges to a harmonious society.* Cambridge: Cambridge UP, 2013. Print.

Shrivastava, A., and Kothari, A., *Churning The Earth: The Making of Global India.* Delhi: Penguin Viking, 2012. Print.

Strathern, Marilyn. *After Nature: English Kinship in the Late Twentieth Century.* Cambridge: Cambridge UP, 1992. Print.

Surviving Progress. Dirs. H. Crooks and M. Roy. Canada: Cinemaginarie and Big Picture Film Corporation, 2011 (DVD release date 2012), 86 minutes. DVD.

Taylor, C. *Modern Social Imaginaries.* Durham; London: Duke UP, 2004. Print.

Van den Bergh, J. "Environment versus growth—A criticism of "degrowth" and a plea for "a-growth"." *Ecological Economics.* 70.5 (2011): 881–890. Web. http://www.sciencedirect.com/science/article/pii/S0921800910004209

Wei, Y.D. *Regional Development in China: States, Globalization and Inequality.* Abingdon, UK: Taylor & Francis, 2013. Print.

Worster, D. *Rivers of empire: Water, aridity, and the growth of the American West.* Oxford: Oxford UP, 1985. Print.

Wright, R. *A Short History of Progress.* Cambridge, MA: Da Capo Press, 2005. Print.

PART III

Tree, River, Ethics

What is "A" Fig?

Nirmal Selvamony

The sycamore fig (*Ficus sycomorus*) in *The Queen of Trees* (2005), the protagonist of the film, has stood on the banks of a river in Tsavo West, Southern Kenya, for the past hundred years. Her co-evolutionary relationship with her pollinator, a tiny fig wasp, has been filmed by Mark Deeble and Victoria Stone in a stunningly captivating manner.[1] Even such people as the Kikuyu and the Masai, who have enjoyed an enduring relationship with this tree, could not have seen or known the exact nature of the relationship between the two organisms because it is not visible to the naked eye. The state-of-the-art technology of the movie enables us to peer into the tiny fruit and observe the journey of the wasp (not more than a millimetre long) into the fruit, and even the rhythmic movements of the nematodes breaking free from her body, ready to lay her eggs before she dies. Another rare opportunity for the viewer is being able to squeeze into the tree-hole and see the female African grey hornbill and her three chicks, even the one who was attacked in the nest by some honeybees. In short, the movie allows us to see both the tree and its inside like never before.

The inside and outside of the tree may be viewed through the lens of the Tamil concepts, ***akam*** (inside) and ***puRam*** (outside). The former is "private" or domestic space, the arena of sexual performance, and the

N. Selvamony (✉)
Department of English Studies, Central University of Tamil Nadu, Neelakkudi Campus, Kangalanchery Post, Tiruvarur District, Tamil Nadu 610101, India

R.K. Alex, S.S. Deborah (eds.), *Ecodocumentaries*,
DOI 10.1057/978-1-137-56224-1_8

latter, "public" or social space, wherein all actions other than the reproductive ones are performed. The true interior (or *akam*) of the tree is the inside of the fruit, the theatre of sexual intimacy. This interior space is occupied by the wasp and the nematode, both of which become transparent, like glass, thanks to the skilled use of cutting-edge photography. Although the tree hollow is, in a way, the inside of the tree, it ought to be regarded as the outside (or *puRam*) as it is not the arena of sexual performance, but a shelter for friends and guests. It is the home of the hornbill, the barbet, and the honeybees (until the Masai men smoke them out). Although the cavities may be viewed from the outside, advanced photography provides us with views of inside the hollow, which would otherwise be impossible to see.

Most *puRam* relationships are played out on the exterior of the tree and on the ground underneath, and even in locations far beyond. We see more organisms (insects, birds, and animals) on the external surface of the tree: fig wasp, caterpillar, ant, long-horn beetle, fig katydid, pink mantis, parasite, bandit wasp, nymph of Hilda bug, honey bee, fig cicada, spider, stick insect, weaver bird, barbet, African grey hornbill, bulbul, African green pigeon, white-eye, fruit bat, gecko, green snake, tree snake, and black-faced langur. Only the giraffe and the elephant can roam under the tree and take her fruit. The butterfly, the vinegar fly, and the seed bug are the insects seen under the tree, while the skink and the deer pick up what has fallen beneath the tree. In the water under the tree, we find fish and the crocodile, although the latter does not enjoy any direct relationship with the tree, like the fish it traps with its gigantic jaws. Far beyond the tree, the fruit travels in the form of juice in the mouth of the African green pigeon to the bird's nest, where its young ones lie waiting, eager to enjoy this sweet syrup. The fruit bat chooses to take the fruit to a distant location, where it purges part of the fruit, which has the potential to produce a new tree. The fruit also travels in several other forms, in the guts of animals and birds, who will eventually disperse them elsewhere.

A rare spectacle such as this, where the cast is a wide variety of organisms (from micro- to macroscopic) who play their roles on the outside and inside of the tree, ought to reveal the true identity of the tree. We might see the tree as an individual in a riverine ecological community. But is she really an individual? Is there any such thing as "A" fig? What do we mean when we call her a "queen?"

Even if the metaphor of queen does not imply a "queendom," it is inseparably connected with a hierarchical community not unlike a state society in which the organisms are all ranked high and low. But is there any evidence of hierarchy? Does the tree play the role of a queen? Let us begin with the charge of anthropocentrism. To call a tree a queen is to be guilty of anthropomorphism ("prisms: nature"). The assumption behind the charge is that a human and a non-human belong to separate biological categories and therefore speaking of one in terms of another is fallacious. In other words, the basis of the charge is the notion that could be called "ontic purism," already challenged by several scholars (Evernden; Findly; Selvamony, "*tiNai* as Tree").

We might trace the roots of ontic purism to Aristotle and the Jain theory of life-forms. If Aristotle is said to have attributed no sensation to plants (Tompkins and Bird ix), a verse in *marapiyal* (chapter on conventions) of *tolkaappiyam* (earliest grammar text in Tamil) has ruled that grass and trees have only one sense—tactility (III.9.28). Considering the basic primal world view of the bulk of *tolkaappiyam*, which does not hierarchise living beings on the basis of their "knowledge" of the world, the attribution of tactility alone to grass and trees is incompatible with the content of either *tolkaappiyam* or the earliest poetic texts in Tamil (Selvamony, "*tiNai* as Tree"). If the *tolkaappiyam* verse is an expression of the theory held by the Jaina thinkers, it must be regarded as an interpolation. As for Aristotle's hard rationalistic view of plants, it is not supported by modern research. Raoul France, a Viennese biologist, avers that plants "are capable of intent: they can stretch toward, or seek out, what they want in ways as mysterious as the most fantastic creations of romance" (Tompkins and Bird xi).

As intent is not possible without consciousness, it must be acknowledged that plants are conscious beings. If consciousness, as Husserl tells us, is object oriented, then any conscious being is conscious of something (Husserl). We might extend Husserl's idea and say that consciousness is value oriented. The relationship between value-oriented consciousness and the world is *a priori*. The tree is already conscious of the other members of the community (*tiNai*) in a value-oriented way. The fig wasp, the spousal partner of the sycamore fig, is the most intimate member in this community. These partners are already conscious of each other's presence in an axiological manner before they consummate their relationship. The plant member is "capable of picking up messages of intent, benign or malicious, that are inherently more truthful than when translated into words" (Tompkins and Bird 32). This means that plants are not only conscious of the presence of other beings, but also capable of sensing the ethicality of the consciousness of their patients and reacting accordingly. The reactions are not mere functions but actions.

If the tree is what she does, then we need to understand her actions. Let us consider the action of latex production. The latex will, hopefully, keep off such leaf-eaters as the fig katydid and the longhorn beetle, who consume leaves, which are necessary for photosynthesis. But the latex does not manage to successfully deter the leaf-eaters who circumvent the problem ingeniously and continue to eat the leaves. This does not mean that the tree is stupid and the leaf-eating insects clever. Both the tree and the leaf-eaters do what their ancestors did and such traditional behaviour, which may be modified to solve new problems, is a vital part of the context in the relationship between the tree and the other beings.

Very often, "relation" is represented as "dependence" or "interdependence" (Selvamony, "Interrelatedness"). The word "depend" hides much more than what it reveals. When we say that the tree, for her propagation, completely depends on a strange visitor, who should arrive one day, no one knows from where, in order to pollinate her flowers,[2] what does "depend" mean? A book on ecology cannot do without this and other similar words. But no such book reveals the true meaning of these words. Is dependence wholly a matter of material transaction such as the intake of nutrition, or the transfer of chemicals? Can we wholly rule out non-material interaction? Are not feelings and values involved in these transactions? Can we exorcise the meaning of "hope" from the word "depend"? How does the wasp or the tree feel about their first meeting? Does the tree "know" that a wasp will visit her? Will not such a crucial event raise expectations on the part of both partners involved? Now, what if the wasp does not visit the tree? If the tree were a human, we would have said that the tree was "hoping" to find her pollinator, unless she decided to remain a "spinster" by producing the right chemicals to keep her pollinators at bay. It is as if the tree called and the wasp responded. From miles away, the elephant also smells the scent of the fig and responds eagerly, although the two responses, the wasp's and the elephant's, are not quite the same. How do we explain this call and response? The complementarity of the relationship involves formulaic pairs: fig–fig wasp, fig–fig katydid, fig–fig cicada, and fig–fig seed bug. Explanation of such a relationship in mere quantitative or physical terms amounts to crude oversimplification. This is not just a matter of food or economy.

The relationship between the tree and her visitors may not even be fully accounted for by means of quantitative analysis. She draws the bee from afar with her (scented) milky latex, which ultimately turns out to be the latter's death trap. Another visitor, namely the fig cicada, ends up a prey to a hunting hornbill. The female fig wasp might have been very happy to enter a new world outside the fruit, when she was helped by the male wasp to break free of the fruit. But did she realise that she was, in fact, about to fly straight into the web of a spider lying in wait for her, or perhaps about to meet a hungry tiger beetle.

In order to understand the actions mentioned above, and others, we may well begin with the nature of action itself. Here, we may turn to *tolkaappiyam*, which gives us a definition of action as that which involves three factors: agent, patient (recipient of action), and context. As the agent is a conscious being, her intention is something we need to factor in, besides the instrument of action, if it is used at all.[3] Context includes place-time and also the ultimate value of the action. It must be pointed out that context is common to both the agent and the patient (*tolkaappiyam* II.3.29).

Although each action will involve all these constituents, the representation of the action itself may not make all the constituents visible. Now, the movie is also a series of actions. All these actions go to make a visual text (with a duration of 52 minutes and 20 seconds) consisting of a single visual sentence which can be analysed into various clauses and phrases. This chapter does not propose to undertake a thorough cinematographic analysis of the visual text, but to deploy formal analytical remarks, especially on action, to bolster this new insight into the ontology of the tree and (indirectly) the human.

Visually represented actions are governed by their own syntax, which is all about sequence and position. If sequence is a temporal category, position is a spatial one. The movie attempts to present a universal truth, namely co-evolution, rather than a momentary truth, which may or may not be always true. The universal truth is conveyed inductively by apparent focus on a single tree. What is true of a single tree ought to be true of all other trees of the same species. In other words, the particular has to yield a universal truth. To this end, the camera affords both sequences, from "one to many" as well as from "many to one." But this is done as if it were a conventional technique of spotlighting a protagonist or establishing the setting of the protagonist of a play through a shift from a panoramic view to a close-up.

If the transition from one visual element to the other is a matter of visual syntax, how helpful are the notions of paradigm and syntagm? A paradigm allows us to substitute one agent for another. But can we say that all actions are random and their agents substitutable? True, there are hundreds of trees and millions of wasps. But the meeting of one particular wasp and a fruit cannot be explained away by the notion of paradigm. All that the paradigm can tell us is that a particular action requires an agent and that there are several who could and do perform the action in question. The agents who are not involved in the action in question are not doomed to inaction; in fact, each agent has her own universe of action. Therefore, the notion of paradigm could exist only in thought and not in reality. An actionless tree or an actionless fig wasp is only a cognitive *figment*. Therefore, there is no paradigmatic axis for any action. Apparently, the idea of paradigm itself is an offshoot of dualistic thinking.

What we have in reality is action (here termed praxis) performed in a particular location. A single praxis, not unlike a sentence, is constituted by one or more significant actions we might call "praxemes," which may be the equivalent of clauses[4] in an unalterable sequence in time. There is no synchronic action. Synchrony is possible only as thought. Like the paradigm, synchrony is a figment born of dualistic thinking. Action is an event in a particular place. For example, a hornbill feeding its young is a praxis that consists of such praxemes as the parent bringing food to the nest, the chick approaching the parent to get that food, and the parent transferring the food to the chick. One of the significant aspects of praxiological analysis is the mode of transition from one praxeme to another. Three such modes occur in the movie:

1. The organismic agent transitions from one praxeme to the next without meeting with any accident that might prevent such a transition. The two chicks of the hornbill, who were lucky enough to fly safely out of the nest, exemplify this possibility.
2. The agent may meet with an accident, thus preventing it from performing the next act. Examples of this could be the third chick who was attacked by the bees, and the cicadas picked up by the hornbill.
3. The agent may meet with an accident but escape unhurt, partially hurt, or gravely hurt and hopes to perform its next praxeme. The first of these three possibilities is illustrated by the wasps who flew away freely without being trapped in the spider web or picked up by predators, such as ants, geckos, and birds; the second, by the gecko who escaped with only a missing tail, which can regrown; and the third, by the butterflies caught by the monkeys and released after a while.

Discussion of the visual representation of the praxis and praxemes cannot ignore context. It is here that we need to comment on the modification of the context by means of music. Although we do have opportunities to listen to the natural sounds of the actions represented, we are also compelled to modify the context of many of the actions with studio music. Such modification changes the very nature of the action. Introduction of music from a context different from that of the action the viewer is watching denatures praxis and Nature, renders the movie gimmicky, and undermines the message. However, from the movie-makers' point of view, they need to put the viewer in the right frame of mind and the music they have chosen, they believe, achieves this end, as in the case of the incidental music in the scene where the butterflies "drink alcohol through straw." If the gain is melodrama, the loss is the nature of Nature. In order to avoid paying such a huge price for a paltry gain, a better option would have been the predominant use of natural sounds and occasional soundtracks of the traditional, indigenous music of the Masai and Kikuyu, especially those associated with the rituals involving the tree. Though avoidance of contemporary tribal music adulterated with western instruments and harmony would have been a challenge to Deeble and Stone when they were making this movie, the choice of music is crucial to the very nature of the movie as it modifies the context of the praxis of the agents in their plot. By lifting the praxis out of its native context, music could disrupt the syntax of the visual narrative.

The praxemic syntactic possibilities we have considered are not exhaustive. But even the ones we have observed involve transitioning from one to the other, not mechanically but holistically. Holistic praxis cannot dismiss the spiritual dimension even on grounds of scientific reasoning, especially if we want to understand the organisms, if not fully, at least, better. We cannot rule out organismic consciousness of actions and attitudes of organisms to those actions. As we pointed out earlier, consciousness is already influenced by the world. It is also tinged with emotion. There is hope in every breast. But no organism, including the human, can read the future beforehand in order to plan the next move, as in a game of chess. If we call our attitude "hope," it is much the same thing with the other organisms too! The agent that hopes for the best also chooses the best possible action at hand. In the case of the fig wasp, she has more than one option with regard to the actions she can perform. Though she is essentially related to the fig fruit, she could enter into other relations as well, especially those that prevent her from effecting the essential one. For example, when the wasp becomes food for the spider, she is unable to effect the essential relationship. This means that every wasp is not essentially related to the tree. Though the wasp alone enjoys an essential propagative relationship with the tree, every fig does not become a patient of that relationship. Why do only some wasps effect the essential relationship? Apparently, there are more fig wasps than necessary for the task and each and every fruit produced by the tree need not be pollinated to propagate the species. If the latter happened, a dominant single species could become a major threat to biodiversity. So, while some wasps accomplish the essential relationship, others become food for various organisms. Evidently, the wasps are both pollinators as well as food. If so, do the wasps have a choice, although both choices are, in the end, fatal? Does the wasp who enters the fig already know that she will eventually die inside the fruit? Does she know that the nematodes inside her body will kill her? Does the wasp make the choice of pollination with knowledge of the risks involved? If so, her choice can be, as Deeble and Stone put it, the "ultimate sacrifice."

We might wish to explain such an act by means of the biological category called instinct. We often understand instinct as a pre-programmed faculty that enables the organism to perform certain acts. But this does not explain it adequately. Instinct may be just another word for tradition, at the level of the gene. Even as humans do things which have been done over and over again by their ancestors, establishing what we have called "context," the non-humans also perform such actions. A hornbill chick is able to fly out of the nest for the first time confidently undeterred by the heights because it is an ability proven by its ancestors. The bird probably has the ability in the form of an instinct but it needs the will to demonstrate that ability. Like us humans, other organisms need to will their acts and will implies choice. This means that humans may not be the only ones who choose their acts. If we think that humans alone have free will and perform actions of choice, we may be proved wrong by several examples of "playful" acts on the part of other beings. The monkeys catch the butterfly or the honeybee knowing full well that it is a "purposeless" act. Therefore, attributing all the actions of other beings to instinct may only be evidence of our own naivety and ignorance. The actions of the agent and the patient are all relevant and purposeful because they are validated by a context shared by both. In order to understand the relevance and purposefulness of the actions, we may do well to distinguish the modes of interrelation between the agents and the patients.

We have already noted that the most important relationship in the movie, according to its makers, is the one between the tree and the fig wasp (Attenborough 140–44). If the scent of the fruit does not attract a straying fig wasp, there will be no propagation. For this reason, the relationship between these two partners is essential. The fig fruit and the wasp are an inextricable pair, shall we say, not unlike spouses.[5] Such a relationship, termed *iNai* (literally, "union"; *cilappatikaaram* 8: 33–34; cuntaram 152–80; Selvamony, "Water…"), meaning spousal, is also the name of the fifth relation in music. Though the pitches C and G are different ones, they sound almost indistinguishable when heard together. Similarly, the wasp and the fruit are quite unlike each other, especially in size, a contrast that is presented brilliantly by showing the fig wasp (small enough to drown in a dew drop) on the antenna of a grasshopper, using the grasshopper as a scale. Such unlike things become a single organism, especially when the wasp is inside the fig fruit, and this is demonstrated through microscopic photography.

Another example of an essential relationship may be the one between the seed bug and the tree. If the seed bugs do not eat up the seeds, several new plants may germinate under the tree. As this is not an ideal site for the growth of new plants, the role played by the seed bugs is an essential one given the fact that no organism, except the seed bug, can remove the seeds from underneath the tree. In this regard, the existence of the seed bug and that of the tree are inseparably related. If so, we have more than one patient enjoying an *iNai* relationship with the tree.

Yet another relationship that may be considered essential is the one between the primal people and the tree. Significantly, the film-makers marginalise this relationship for obvious reasons. At the beginning of the film, they refer to the myths and legends of the tribal people in just one line, and later, they show the Masai making fire from the wood of the tree and smoking out the bees inside the bark to gather honey. They carefully avoid anything that might be considered "spiritual," as that might go against the scientific approach of the film. The film does focus on a very important aspect of the tree—co-evolution. But the queen is much more than just her affair with the wasp. She is none other than the ancestor to the Kikuyu tribe who pray to her, according to Wilde, for wealth and rain. We also learn that their women, who seek the boon of children, smear themselves with the latex of the tree. The tree is believed to ensure fertility in other animals too. In the hope of augmenting their virility, men sleep on the leaves of the tree ("Queen of Africa's Trees"). The beliefs of the people constitute the context that gives meaning to the praxiological relationship between the tree and the people. Evidently, an ancestral tree is an essential part of one's identity.

Besides the kind of relationship in which partners achieve spousal identification, there are others in which they stand in kin relationship with each other. Consider, for example, dispersal of seeds, an action performed by several birds and animals that consume the fruit and leave the seeds in their droppings. For the elephant and the deer, the fig is not the only source of nutrition. Other animals could also have consumed the fruit and dispersed the seeds. The agent is substitutable, but not the act of dispersal. This does not mean that this is a paradigmatic relationship because the dispersing agent is not selected from a group of potential dispersers. We are only saying that there are several agents who could perform this particular task, though only one performs at a time. This relationship is called *kiLai* (literally, "branch"), meaning "filiative" or kin-like, not unlike a sibling. The seed-dispersing agents enjoy kin relation with the tree and the relationship between the two may be described as "partial identification." This relationship is like the fourth in music, wherein the two pitches do not sound as if they are one, but blend to manifest identity and difference, even as the branches of a tree do; they are different from each other and yet identical with the tree. Different branches, like the *kiLai* relation, perform more or less the same biological actions and, therefore, are replaceable.

If the actions of *iNai* and *kiLai* are essential to the tree, there are also others that are not necessarily essential but important. The role of the ants guarding the "secret garden" (fig fruit) is one such important action. For this reason, Deeble and Stone call the ants "allies." The fruit in which the wasp chooses to enter and pollinate has to be guarded for at least two weeks, until the eggs of the wasp hatch. Guarding the pollinated fruit is essential only if that particular fruit is likely to be consumed or destroyed. But this is a possibility rather than a certainty. Therefore, the action of the ant can only be deemed as an important one rather than an essential one, and the agent who performs such an action may be considered *naTpu* (friend), and the relationship "affable" (which is also denoted by the term *naTpu*). In musicology, *naTpu* is the third relation, as the one between, for example, C and E natural. The tonal agreement between a pitch and its third is not identification but a "sharing of some commonness," even as friends are bonded by common interests (Selvamony, "Water ..." 91).

The fig not only has friends but foes as well. The parasites,[6] the bandit wasp and nematodes, are the biological antagonists of the tree. Like the second relation in music, *pakai* (antagonism), which effects neither identification of pitches nor commonness among them, but only discord, the antagonists play a significant role in a community. Though they are the voices of dissent, in the final analysis, they ensure the common good.

To these relations, we may add one more not unlike the "ornamental" one in music. For example, the natural third (E) and natural seventh (B) are ornamental pitches in the musical mode now known as *aanantapairavi*, which are not part of the ascent and descent of the mode but used rarely as grace notes (cellatturai 134). We may call such pitches *viruntu*, guests who are not part of the kin group (like *iNai* and *kiLai*) at all but important members of any primal home (*tolkaappiyam* III.4.11; *tirukkuRaL* 43). They help nurture the value of hospitality. Consider the relationship between the fig tree and the honeybees. The fig flowers inside the fruit neither see the sun nor are they visited by the bees. Only the fig wasp can pollinate the microscopic flowers. The bees could have made any other tree their home. In other words, neither the bee nor its action of building the hive is essential to the host tree. From a biological perspective, the relationship between the two is neither essential, like *iNai* and *kiLai*, nor important (*naTpu*), nor in any way antagonistic (*pakai*), but contingent. But from a spiritual perspective, such an ornamental relationship is necessary.

What about the relationship between the fish and the tree? If providing nutrition to several organisms is a function of the tree, the fish and the tree stand in a trophic relationship. However, fig fruit is not a normal nutritional source for fish. In other words, the trophic function is not an essential one as far as the fish are concerned. It can only be an optional function. Yet, the context, namely the aquatic habitat underneath the tree, is not an optional one but a given one, which makes the relationship complex. Underneath the tree we have two habitats: land and water. Those fruits that fall on the land bring one set of agents to the tree, while those that drop into the water bring another. Due to the presence of the aquatic habitat under the tree, the migrating catfish is able to take the fruit with it. This relationship between the catfish and the tree, and also that between the other fish (which enjoy the fruits dropped by the monkeys) and the tree, is contingent and, therefore, *viruntu*.

So far we have shown that the fig tree is related to other organisms in various ways, as *iNai*, *kiLai*, *naTpu*, *pakai*, and *viruntu*. What do all these different relationships mean? Do they give us any insight into the nature of the tree? In the *iNai* relationship, a particular patient (the wasp) is essential for pollination, although other patients are also involved.

The other agents who could be substituted for purposes of pollination do not stand in an *iNai* relationship but in a *kiLai* one. If there are allies (*naTpu*) to the project of pollination, there are also opposers (*pakai*) and others who do not seem to be connected with it in any significant way (*viruntu*).

From a broader perspective, relations other than the essential ones are also equally important and necessary. We seem to think that the propagative relationship between the tree and the wasp is essential, whereas the trophic relationships (between the tree and all the organisms she feeds) are not essential. We might argue that without propagation, the tree will become extinct, but she can still survive without feeding other organisms. In other words, our parameter for valuing one relationship higher than another is the principle of life. Such valuation presupposes that life is an ultimate value. Unlike contemporary society, *tiNai* society did not regard life as the highest value (*tolkaappiyam* III.3.22:1–2). If quality of life (as in primal societies) is our parameter to evaluate relations, non-propagative relations will also assume importance. The trophic relation contributes to the quality of ecology, especially by ensuring biodiversity. Quality derives its importance more from the context of action than from the agent or patient.

Without the context, we have no way of understanding what kind of relationship obtains between the agent and the patient. Each relationship involves a specific type of agent—spouse(s), kinfolk, friend, guest, and antagonist. No primal home is complete without all of these agents. Even antagonists play a significant role. They are the deterring agents (Miller) who check and oppose the actions of the family members when necessary in order to strengthen genuine relationships and ensure the ultimate good of the family. Examples in the early Indian home (called *tiNai*) are the counsellors (*aRivar*, literally, "the ones who know"), who even have the right to reprimand family members. However, excessive antagonism could result in violence and destruction. It must be noted that kinship is not the only relationship among the members of a primal home.

The primal home, known as *tiNai*, includes all the different stages of heterosexual relations between the couple. There are five basic modes of relationship and each one is associated with a typical land area (*tolkaappiyam* III.1). If the lovers meet in the mountains and fall in love with each other, they make their home in the scrub jungle. When the man marries a second woman, a practice which enjoys social acceptance, tiffs are common and such domesticity is located in the riverine plains. Routine separations, like overnight fishing, are set in the coastal areas, whereas occasional longer departures from home for the sake of finding wealth, political assignment, or learning are associated with arid places.

Each of the five modes of relationship may be explicated with the help of the modes of tonal relationship. When the lovers first meet, they are strangers until they fall in love. The first union is mental, predominantly emotive. In fact, it is a complex one, at once emotive, volitional, and cognitive. This complex union, which is not yet physical, is said to be a union of hearts. Metaphorically, the heart stands for the multiple mental faculties. *kuRuntokai* 40, an ancient Tamil song, written by cempulappeyalniiraar, conveys this stage of the relationship quite effectively, in its last two lines: Like rain water and red earth,/our hearts in love have united (Trans. Nirmal Selvamony).

When the union is only mental, although it is a union of multiple faculties, it is considered unilateral, such as that which is found in friendship (*naTpu*). Shakespeare called such a union the "marriage of true minds." It is also comparable to what is referred to as "platonic love." The ants that guard the pollinated fruit enjoy friendly relations with the tree, although such a relationship is not a holistic one, such as that of the pollinating fig wasp and the tree.

If the relationship between the partners is only mental and not physical, it may not be a holistic one. In the case of the sycamore fig, the ants are always allies; they never become sexual partners like the fig wasps. However, we do not see any tangible evidence for the transformation of the friendly fig wasp into a sexual partner. Probably, we need to know all about the entire pollination project of the wasp to be able to tell anything conclusive about its pre-sexual contact with the tree.

In human terms, the domestic life of the wasp is quite brief. If there is anything like the equivalent of home-making, it is what happens between the wasp landing on the fruit and dying inside. The pleasures of outdoor life which can be enjoyed from indoors in a scrub jungle—the rain, the fragrance of flowers, and the changing colours of life-forms—are not within the reach of a fig wasp. Yet, the fig tree and the wasp enjoy the closest intimacy possible, which is the equivalent of what we have termed an *iNai* relationship.

Though the pollinating wasp alone enjoys greatest intimacy with the tree, there are others, not unlike subsequent wives, whose relationship with the tree is also physical. For example, the parasites and the banded wasp do not prevent the fig wasp from pollinating the tree but introduce checks and balances in the inter-member relationship in the community, not unlike the *kaamakkizatti* (the legitimate second wife in a love-based primal *tiNai* home). Such checks and balances do not exclude the killing of the wasp. Excessive deterrence, personified by such a persona as the courtesan, brings pleasure and ruin at once. It may be possible to compare the teleology of antagonism with that of elusive evolution itself. The invisible hand of evolution, shall we say, encourages the tree to check the population of the parasites by encouraging their predators, such as the ants.

Besides the friendly ant and kindred parasites, there are also antagonistic nematodes, which kill the spousal wasp. Other antagonists are the leaf-eaters, on account of their draining the tree of her much-needed energy to sustain her fruits and wasps. Such antagonistic relations (*pakai*), on the face of it, look evil, but from an evolutionary perspective, they may be legitimised as part of the holistic fabric of the home, called *tiNai*. From this perspective, we might say that this is "necessary antagonism." Finding wealth to support the family means temporary separation on the part of a partner, usually the male, when the female is left behind to sustain the home. Education, and social responsibilities and assignments separate the partners, at least temporarily, and in this respect, they may be counted as necessary antagonistic forces in a *tiNai*. Similarly, the leaf-eaters, fruit-eaters, and seed-eaters are not to be painted as pure villains in a monolithic manner. They may be seen as agents who test the strength, endurance, and dynamism of the tree and encourage the latter to fare better.

So far we have considered the inner (*akam*) life of the tree through the phases of the meeting of the partners, and consummating and deterring sexual union temporarily and permanently. But the tree also lives its own outer or public (*puRam*) life, as the greatest provider in its habitat. Responding to the various needs of each member of the habitat, the sycamore fig is not so much a member of *puRam tiNai*, who performs the role of a provider, as that of *tiNai* itself by virtue of being a household of natural and supernatural agents. According to *tiNai* theory, providing sustenance to dependents is the *tiNai* called *paaTaaN* (literally, the sung ruler).

Though the tree is a laudable inhabitant (shall we say, a singable queen) of the riverine habitat of Kenya, she does not expect her beneficiaries to sing her praises. She gives to them unconditionally (tagkappaa 14–15). The hornbill uses her hollow as its home, the honeybees make their hive insider her trunk, the woodpecker finds its food on the tree bark, the monkeys, birds, and bats can take her fruit even from her crown, while the elephants and giraffes have their fill without even needing to climb her. Other animals, such as the deer, pick up what falls to the ground. We do not know how these creatures entice the tree to yield her bounty. Considering the fact that plants have the capacity to sense the intention of their patients (Tompkins and Bird), they could probably differentiate between those beneficiaries who deserve their generosity and those who do not, and if they could, they could also make a difference in the quality of the gift they make.

If the tree is *tiNai*, *tiNai* itself is the tree. There are three basic constituents of any *tiNai*, the base (*mutal*), the generative (*karu*), and the praxiological (*uripporuL*). The base of the tree is called *mutal* because it bears the entire weight of the tree's body and also provides the necessary food for all the sub-systems of the tree through its root system. The large body is supported by timber, the dead tissue, which is found abundantly at the base rather than anywhere else (Attenborough 51). The root not only locates the tree in a particular place, but also moves about to some extent to find necessary sources of energy. Without the base, the tree cannot be where it is and undergo the necessary changes it has to make. In other words, the base determines the spatio-temporality of the tree. Unlike the other organisms, the tree has the most determinate location. This is especially true of other species of *Ficus*, such as *Ficus benghalensis*, which has numerous lateral roots that establish the tree very firmly. In fact, the very idea of stability or emplacement is metaphorised by the tree. No word can convey this sense better than "rootedness." Therefore, no other organism can convey the *a priori* nature of place-time such as the tree can.

Like any other organism, the tree also produces a variety of entities. Through its root system, it produces food from the ground resources, and by means of its leaves, from sunlight. Although the tree consumes food from the earth, it is not regarded as a consumer because it does not depend on any organism other than itself for its own food. True, trees do depend on other creatures for pollination. But they produce their own food, and for this reason, they are called producers. If the herbivores are direct consumers of the tree, the carnivores are indirect consumers, as their food is the herbivore prey (Attenborough 54). A third mode of production is reproduction. As an organism that produces for itself and others and also reproduces, the tree is producer par excellence. Therefore, the generative aspect (*karu*) of *tiNai* (home) life is best exemplified by the tree.

The tree is able to perform all its actions, feeding, growing, reproduction, and so on, because it is well protected by a strong bark (*uri*). Besides the intra-dendric functions mentioned above, there are also inter-dendric ones. It is the bark that gives the tree its form and identity, and differentiates it from the rest of the habitat. Differentiation is complemented by continuity with the rest of the habitat. This continuity is visible when the bark becomes a pathway for creepers, or a hideout for worms until the woodpeckers find them. More intangible functions, such as control of temperature, maintenance of immunity, absorption of moisture, and so on, also establish the tree's continuity with the other members of its habitat (Puplett; "Structure and Function of the Skin").

In relation to the other organisms of its habitat, the tree is one of the members of its community (*tiNai*). The community of the sycamore fig is quite a large one. From microorganisms to wasps one millimetre in length, and macroorganisms the size of an elephant, the community of the fig is almost like an extended family, including the spirit beings that dwell in it and the Masais who regard her as their ancestor. On account of the kinship between the tree and the humans (Durkheim 124; Selvamony "*tiNai* as Tree" 227–29), the tree is not merely a non-human individual (Evernden 95) but also an interspecific agent who shares the traits and qualities of the community (*tiNai*) to which she belongs.

The agents who make up the primal home are not autonomous beings or individuals. For one thing, they are interdependent on each other for performing their essential functions. Further, the being of one entity is defined by the being of another. The being of the tree is defined by the being of the wasp, as much as the being of the wasp is defined by the being of the tree. The tree is at once differentiated from, and continuous with, the fig wasp, even as the arm is in relation to the body. An interspecific being, such as the tree or the wasp, is not so much, as ecological science tells us,[7] an individual as a member of a community whose identity can be defined only in relation to the other members. She is a community (*tiNai*) herself to which several other members belong. If so, how can fig be "a" fig?

Notes

1. In fact, "every species of fig has its own species of fig wasp" (Scott 103).
2. One is tempted to compare and contrast the relationship between the tree and the wasp with that between the human lovers as described in *akapporuL* grammar in Tamil tradition.
3. We may note that tolkaappiyar's theory of action states that intention is a part of any action. If action is a conscious event, then, it follows that according to this theory, consciousness is already intentional. This theory predates

Husserl's by several centuries. "The central doctrine of Husserl's phenomenology is the thesis that consciousness is intentional, a doctrine that is borrowed from Franz Brentano. That is, every act of consciousness is directed at some object or other, perhaps a material object, perhaps an 'ideal' object—as in mathematics." (Husserl)

4. This word is coined on the analogy of morpheme: praxis, action + [s]eme, literally, sign, here, significant; praxeme, significant action.
5. Ultimately, the male pollen and the female ovule are the two heterosexual partners in the relationship we describe. But these two cannot come together without the agents, the wasp and the fruit of the tree. In the case of the wasp, it is always a female who spreads the pollen on the stigma of the fig flowers (Attenborough 141). In other words, a female wasp bears the male part of the flower (namely, the pollen) and performs the masculine role in the reproductive drama. But the case of the fig fruit is complex. Although it has both male and female flowers, by offering the female parts, namely the stigma and the ovule to the wasp, the fruit performs the feminine reproductive role. Ultimately, it is a heterosexual union effected by the agents, namely the tree and the wasp.

   Further, the term *iNai*, meaning "spousal," is used for musical pitches in ancient Indian sources. One may wonder how this term could be applicable to a tree and an insect that are not members of the same set. In fact, our notion of a set itself stands in need of revision. For example, the relationship between a leaf insect and the tree that produces that leaf is much closer than that between one tree and another. In other words, entities which do not belong to the same set (insect and tree) are ontically much closer than entities that belong to the same set (two trees).

   Yet another point to note is the need for new terminology. One may consider this unwarranted. The question may be: "When we have terms such as symbiosis, commensalism, mutualism and parasitism, why go for musicological terms to describe interrelationship?" Though the existing terms pertain to two unlike species, the relationship they describe is only energy oriented or economic. The anthropic-sounding terms "harm," "usefulness," and "help" also have to do with finding energy or survival. The non-material (or spiritual) dimension of the interrelationship is not within the purview of these terms.
6. While parasites are generally regarded as the biological antagonists, there is also the view that they are unnecessarily vilified and deserve more objective treatment. In the humanities, this ambivalence is theorised by Hillis Miller, who argues that "it is impossible to decide which element is parasite and which host…" (Miller 459).
7. The most fundamental unit of "ecology" has to be the community rather than the individual. In fact, there is no such thing as the individual; it is but a figment of our imagination. All organisms are already always members of communities. Therefore, autecology seems to be the most "unnatural" pursuit.

References

Attenborough, David. *The Private Life of Plants: A Natural History of Plant Behaviour*. London: BBC Books, 1995. Print.

cellatturai. pii.Tii. *tennaka icaiyiyal*. (South Indian Musicology). tiNTukkal: viTiveLLip patippakam (viTiveLLi Publishing), 1995. Print.

cilappatikaaram, ed. u.vee.caaminaataiyar. tajcaavuur: tamizp palkalaik kazakam (Tamil University), 1985. Print.

cuntaram. vii.pa.kaa. *pazantamiz ilakkiyattil icaiyiyal* (Musicology in Ancient Tamil Literature). cennai: caiva cittaanta nuuRpatippuk kazakam (Forum for Publishing Works on Saivism), 1986. Print.

Durkheim, Emile. *The Elementary Forms of Religious Life*. New York: The Free Press; London: Collier Macmillan Publishers, 1965. Print.

Evernden, Neil. "Beyond Ecology: Self, Place and the Pathetic Fallacy." *The Ecocriticism Reader: Landmarks in Literary Ecology*. Ed. Cheryll Glotfelty and Harold Fromm. Athens, Georgia: U of Georgia P, 1996. 92–104. Print.

Findly, Ellison Banks. *Plant Lives: Borderline Beings in Indian Traditions*. Delhi: Motilal Banrsidass Publishers Private Limited, 2008. Print.

Husserl, Edmund. "Phenomenology – Edmund Husserl – JRank." Web. 24 April 2016. <science.jrank.org/…/Phenomenology—Edmund-Husserl.html>

kuRuntokai. Commentator. tamizaNNal. Ed. tamizaNNal and cupa.aNNaamalai. kooviluur: kooviluur maTaalaiyam, 2002. Print.

Linford, Jennifer. *Trees*. Parragon, 2007. Print.

Miller, Hillis. "The Critic as Host." *Critical Theory Since 1965*. Ed. Hazard Adams and Leroy Searle. Tallahassee: UP of Florida, Florida State UP, 4th printing, 1992. Print.

Merriam-Webster Dictionary. Web. 14 April 2016. <www.merriam-webster.com/dictionary/platonic%20love>

PRISMS: "Nature. The Queen of Trees. African Queen." Web. www.prisms.mmsa.org/review.php?rid=1050

Puplett, Dan. "Tree Bark." Web. 12 April 2016. <www.treesforlife.org.uk/forest/forest-ecology/tree-bark/

The Queen of Trees. Dir. Mark Deeble and Victoria Stone. BBC, Granada International, 2005. Film.

Scott, Michael. *Ecology*. Oxford: Oxford UP, 1994. Print.

Selvamony, Nirmal. "Interrelatedness: *tiNai*, Ecocriticism and Envirocriticism." *Gandhigram Literary Review* 1.1 (Summer 2012): 61–80. Print.

Selvamony, Nirmal. "*tiNai* as Tree: Revisiting Tree Worship in Tamil *tiNai* Societies." *Ecology and Life Writing*. Ed. Alfred Hornung and Zhao Baishang. Heidelberg: Universitatverlag, Winter 2013. 215–39. Print.

Selvamony, Nirmal. "Water in Contemporary Tamil Literature." *Words on Water: Literary and Cultural Representations*. Ed. Maureen Devine and Christa Grewe-Volpp. Trier: Wissenschaftlicher Verlag Trier, 2008. 89–104. Print.

"Structure and Function of the Skin—CliniMed." 2 April 2016. Web. <www.clinimed.co.uk/Wound.../Structure-and-Function-of-the-Skin.aspx>

tagkappaa, ma. le. *koTuttalee vaazkkai* (Life is Giving). putuvai: vaanakap patippakam, 2001. Print.

tirukkuRaL (*The Sacred Kural of Tiruvalluva Nayanar*). Ed. Rev. G.U. Pope. New Delhi, Madras: Asian Educational Services, 1990. Print.

tolkaappiya muulam: paaTa veeRupaaTukaL: aaznookkaaiyvu (*tolkaappiyam* original with textual variations from a research perspective). Ed. K.M. Venkataramaiay, S.V. Subramaniam, and P.V. Nagajan. tiruvanantapuram: International School of Dravidian Linguistics, 1996. Print.

Tompkins, Peter, and Christopher Bird. *The Secret Life of Plants*. Rupa & Co., 3rd impression, 2006. Print.

Wilde, Samantha. "Queen of Africa's Trees: The Sacred Fig Tree." 2 April 2016. Web. <www.deeblestone.com/film_info-php?filmld=1&typeld=5>

Elemental as Mythic and Existential Parable

John A. Duvall

Introduction

Shot in direct cinema style (live footage edited without narration to establish maximum realism), the documentary *Elemental* (2013) follows three very different activists through their respective journeys and challenges. What is so strikingly original about *Elemental* is its personal focus on the daily lives of its subjects—their commitment and their daily struggles against seemingly insurmountable odds. The film's audience can see how much they sacrifice for their devotion to their causes, and how deeply those sacrifices are rooted in their own sense of identity and spirituality.

Rajendra Singh is a Commissioner of the Ganges River Basin Authority in India, who embarks on a pilgrimage along the polluted Ganges River in an attempt to raise the consciousness of villagers to inspire greater respect for the sacredness of the river. Along his journey, he encounters opposition from his fellow Indians and his own personal self-doubts.

Eriel Deranger is a young mother of the Dené Athabasca Chipewayan First Nation of Alberta, Canada, as well as an activist for the Rainforest Action Network (RAN). She campaigns against the Alberta tar sands—potentially the world's largest supply of oil reserves—and the proposed

J.A. Duvall (✉)
Dominican University of California, 5311 Daniel Dr. Rohnert Park, San Rafael, CA 94928, USA

R.K. Alex, S.S. Deborah (eds.), *Ecodocumentaries*,
DOI 10.1057/978-1-137-56224-1_9

Keystone XL oil pipeline, to protect her people from threats to their health and the destruction of their traditional lands and ecosystems.

Australian inventor and entrepreneur Jay Harman applies his understanding of biomimicry—the utilisation of natural processes in the design of human technologies—to create inventions that hold out the promise of limiting the environmental damage from climate change. He spends a lot of his time seeking investors who might be willing to risk millions of dollars in support of his designs for strange devices they can barely understand.

The most important focus for the film-makers is not whether these activists are right or wrong in their beliefs, but that they are driven by an existential commitment that overshadows everything else in their lives. *Elemental* engages topics that rarely show up in environmental documentaries: the confusion of regular people towards powerful forces that offer them employment but endanger their health and ecosystems, the divisions within the politics of environmental movements, and the constant tension between public activism and personal life.

The primary approach of this current analysis will be through a lens informed by literary myth, spiritual symbolism, and existentialist philosophy. The film's title suggests a link between the natural elements (in classical terms, earth, water, fire, and air) and the protagonists' personal commitments to heal humanity's relationship to the natural world. Their beliefs are profoundly influenced by their spiritual heritage, their cultural roots, and their personal experiences of nature.

Our discussion will build on various reference sources. The film's website offers a discussion guide explicitly rooted in Joseph Campbell's portrayal of the hero's journey from his book *The Hero with a Thousand Faces*, framing the protagonists' growth from their earliest consciousness of nature into their lives of committed activism. Romanian historian of mythology Mircea Eliade, in his books *The Sacred and the Profane* and *Patterns of Comparative Religion*, studied the symbolism connecting spirit and nature, searching for the roots of religion in the natural world. Prominent philosophers associated with existentialism, from Martin Heidegger to Jean-Paul Sartre, have framed the relationship between existential crisis, personal identity, and nature on a more secular, psychological plane, considering whether a person's decision to struggle against the inevitable may be the best strategy for defining one's identity in society and warding off despair.

While many of these thinkers are regarded as politically conservative in their sometimes romantic elevation of the beliefs and traditions from past eras, their views on the relationships between nature and the sacred, and their observations on modern man's psychological alienation from nature, have served as a fertile soil for the development of an ecological psychology—an important development as mankind faces its greatest environmental challenges ever from pollution, fossil fuel depletion, widespread species extinction, and anthropogenic climate change. However, the most crucial sources of insight come from the self-revealing testimony of the protagonists within the film itself. As they reflect on their personal histories in family, nature, and spiritual tradition, they reveal the deep relationships between their self-identities and their actions in the world.

Joseph Campbell and the Hero's Journey

Joseph Campbell (1904–1987) was one the most esteemed scholars of myth and comparative religion in the second half of the twentieth century—certainly the most publicised, due to his book *The Hero with a Thousand Faces* (which film-maker George Lucas cited as an inspiration for his *Star Wars* films), his four-volume work *The Masks of God*, and his televised series of interviews with journalist Bill Moyers entitled *The Power of Myth*.

The sources of Campbell's ideas may be found in the literature of religion, psychology, philosophy, and fiction. He began his academic career as a student of James Joyce and Thomas Mann, and was greatly influenced by the writings of Sigmund Freud and Carl Jung, especially with regards to Jung's theories about archetypes, dreams, symbolism, and the unconscious (Paraschivescu 217). His interest in myth spanned both Eastern and Western traditions, and his life's work was an attempt to synthesise the varied strands of world mythology into a unified whole (Gentile 2007). Borrowing the term *monomyth* from Joyce's *Finnegan's Wake*, Campbell attempted to demonstrate that the myths of the world's seminal religions and cultures conveyed common themes and structures, which still hold important meanings that may help to heal alienation and benefit individual psychological integration in the modern world (Ellwood 143).

In *The Hero with a Thousand Faces*, Campbell describes several stages of the hero's journey (Campbell 1949; Vogler 1992). The most universal include: the Call to Adventure, in which a shift in the external or internal world thrusts the subject out of his or her comfort zone into a new challenge; the Refusal of the Call, during which the would-be hero vacillates and seeks a way to avoid facing up to this change; Supernatural Aid, often including guidance from a spiritual Mentor; the Road of Trials, in which the Hero endures many rites of passage, crossing dangerous thresholds filled with shadowy presences; the Apotheosis and Ultimate Boon, whereby the Hero faces his supreme danger, overcomes his limitations, and achieves his goal, changing both the world and himself in the process; and the Return, wherein the Hero brings his achievement and his wisdom back to his world and his people.

Elemental's educational discussion guide, prepared for the film's website for high school and college-level students, approaches its protagonists' stories based on Campbell's framework, focusing on three acts: the *Call to Action*, the *Road of Trials*, and *Finding Hope/Coming Home*. Of course, being a documentary film, *Elemental*'s stories do not have neat beginning and ending points, since the beginnings predate the film's production and the endings are yet to be realised. The film builds on Campbell's structure through three themes—*Connection to Place*, *Perseverance*, and *Nature as Teacher*—appealing to the psychology of younger viewers by employing a familiar narrative perspective. The structure of *Elemental* follows its protagonists through these stages of development, demonstrating how these themes play out in their lives (*Elemental*_Discussion_Guide.pdf).

In the film, Rajendra Singh embarks on a pilgrimage to purify the Ganges River, which provides 60 % of Indian's drinking water, to awaken those who live on its banks to both environmental and spiritual dangers. People call Singh "the Rainman of Rajasthan"; he has already succeeded in reviving seven smaller rivers and helped over 1000 villages, but saving the Ganges is a challenge of a different magnitude.

When he visits a town with a polluted lake, an admiring bystander tells him, "Only someone like you can start a war of purification. Nobody else can do it" (00:03:23–00:03:32). The town's mayor expresses support for cleaning the lake, but Singh discovers that the lake has been treated with chlorine in order to confuse the results of chemical tests. This is only one example of the resistance that Singh encounters in confronting people who don't really share his values.

During his pilgrimage, Singh receives criticism from allies and foes alike. Sometimes people react angrily: "Many people like you come here, they make pledges and promises, but nothing happens. This is our tragedy, our pain. How can you help us? What is your solution?" Singh counters by confronting them: "Take a good look at yourself. You live in Haridwar, but what have you done for it? Nothing…I need good friends like you who are angry that nothing's happened yet" (00:37:20–00:38:08). When people in another town complain that he is all talk and no action, he leads them in a physical clean-up operation, where many villagers join him in wading into the Ganges and physically removing trash from the river—a disgusting and dangerous enterprise that literally makes Singh throw up.

A meeting Singh attends between villagers and supporters of a dam construction erupts into an angry shouting match, where Singh is attacked for interference in local matters and for not appreciating people's need for jobs. Singh is caught in the middle between feuding factions that view him either as intrusive or as ineffectual. He says, "I am not a politician. I am a social worker. That is why I was hurt" (00:56:20–00:56:29).

In one of his later travels, Singh hikes up into the Himalayas to the source of the Ganges to visit a hermetic wise man named Mata Bhakti Priyanada, who says, "You yourself cannot do anything. But if it's God's wish, then you will be able to (save the Ganges). This world will be the way it's meant to be" (10:19:19). As well as this: "Good and evil always walk together. If there's no evil, there's no good. There has to be both saints and sinners. No difference then. If we understand that, it is a game. Just play the game" (01:23:59–01:24:20). This encounter with a wise man allows Singh to come to terms with his own limitations and let go of some of his inner turmoil.

In Canada, Alberta's tar sands extraction zone is the largest man-made industrial project on the planet. The face of the land has been transformed by the corporation SynCrude, which removes four tons of dirt for every barrel of oil that it produces. The tar sands must be mixed with highly toxic chemicals to extract the oil, leeching roughly 11 million litres per day into local river systems.

Eriel Deranger self-identifies not as an environmental activist but as an indigenous rights activist. Her strategic goal is to stop the construction of pipelines to carry tar sands oil from Alberta to the western Canadian coast and through the USA to the Gulf of Mexico, via the proposed Keystone XL pipeline. Her family is among those feeling the brunt of environmental destruction; many native peoples downriver from the tar sands are dying too young of cancer, despite living in what appears to be an idyllic location.

Deranger's "call" comes in part from her family heritage; she feels privileged to be born into a family that was politically active. But even though mother and daughter are both activists, there is still a generational divide.

Eriel says an activist has to "learn how to pick your battles...Sometimes it's just not worth the energy. Sometimes you just gotta walk away from people, and [her mother] just couldn't walk away from anything or anyone." But her mother replies: "We had to in my generation, so you could enjoy what you enjoy. The doors that are open now weren't open then. I couldn't even go into a group of white people before without feeling really inferior or that I couldn't speak up or anything" (01:02:06–01:02:31). Her mother, thus, serves her as both wise counsel and critic.

Some of Deranger's "trials" come from her personal life. The mother of one teenage girl with another baby on the way, she must balance her family responsibilities with her passion to protect her people from the overwhelming economic and political power of the energy corporations. She emphasises that her work is not an individual project; it involves organising group protests and engaging in political lobbying. On one occasion, Deranger interrupts an Enbridge Pipeline Advisory Board meeting and confronts employees for violating requests from First Nations communities for Enbridge to refrain from entering indigenous territories, citing a United Nations declaration requiring prior informed consent to do so.

Sometimes, however, Deranger encounters opposition from her own allies. At one point, she attends a fundraiser for her RAN sponsor. She says, "I often get annoyed with rich affluent people that just want to throw their money at people. But I've had to hobnob with donors before" (00:29:43–00:29:59). She experiences an awkward encounter with Robert F. Kennedy Jr., who is rather condescending in giving her advice, saying "Don't screw this up." A representative of her parent organisation, RAN, responds, "Eriel's certainly a controversial person at times. I think she might have spoken a little too loudly about certain people or institutions that had a lot of influence, and our fiscal sponsor in Canada has decided that we're too risky, and they're going to have to cut us off" (00:48:30–00:48:49).

Reacting to her dismissal from RAN, Deranger says, "When I first got involved with the campaign, I definitely had some issues with the structure of RAN, because there was a lot of criticism of the way that I worked for the first year...I'm working with the communities to legitimise this campaign, because without communities buying into this campaign, I'm not going to ask anyone to support it" (00:40:47–00:41:15). Thus, she responds to the setback of her dismissal with renewed determination, and a march that she helps organise against SynCrude in Alberta is a resounding success.

Jay Harman's personal story is less connected to spiritual or cultural roots, and more to his direct contact with nature. He grew up isolated from other human beings and used to run away from his brutal Jesuit school and go down to the beach: "I spent every day that I could in the water swimming, diving. So I started noticing how fish swim, how seaweeds change their shape when waves go past. And so my whole life has become a fascination with movement" (00:10:31–00:10:51). Harman still spends time scuba diving and fishing in the sea.

Today Harman is the founder of PAX Scientific. He has had successes but also suffered setbacks and discouragement along the way, and works with virtually no budget. His current research projects include a water purification device and an atmospheric mixer to cool down the atmosphere and slow global warming. The latter's central idea is to use an aircraft engine with a shell-shaped exhaust cone to push hot air upwards, cooling off an area 50–100 miles wide to 30,000–40,000 feet high. His wife, Francesca Bertone, is his closest ally, and together, they make a powerful team. She complements Jay's otherworldliness by supplying the practical business sense and logistic planning for their company, allowing Jay's work to move forward. Francesca reflects: "He was the kid who… sat in class and was just off in his own world…Lawyers and phone calls and contracts and financial matters and trying to run the business don't suit Jay's brain…I mean that's the visionary, right? The visionary style" (00:32:29–00:33:10).

In the film, Jay makes presentations to potential investors looking for game-changing technologies with millions of dollars of capital to offer, but who have understandable resistance to investing in radical technologies that are difficult to understand. One of his potential investment partners observes, "To bring radical ideas into existence takes a certain amount of investment. But where do you get that funding from? Because the way the world is structured at the moment, there's really no market for early-stage technologies" (00:34:02–00:34:19). But Harman asserts, "We're not doing anything that nature hasn't already done. We're accelerating what nature does" (00:44:20–00:44:27). Harman believes innovative technologies could reduce the world's energy bill by at least half over a couple of decades, if the will is there.

Despite the overarching frustrations and futility of their endeavours, by the film's conclusion, each activist is able to claim a small victory. Rajendra Singh is granted a meeting with the prime minister and gains promises that no dams will be constructed in the upper region of the Ganges.

One hundred and sixty-five industries that have dumped their wastes into the Ganges have been shut down. He continues to travel all 210 tributaries of the Ganges, mobilising communities to take responsibility to protect India's rivers. Eriel Deranger's activism has helped draw international attention to the issues of the tar sands and its projected pipelines. Over 10,000 protesters surround the White House to protest plans for the Keystone XL pipeline. Jay Harman's atmospheric mixer is now moving towards full-scale field trials, and his water purification technology has been proven and is being brought to market. Thus, *Elemental* completes the story arc of each of its "heroes," from their core inspirations and purposes, through their trials with foes, friends, and self, to arrive at an end goal that represents at least some degree of achievement and success. Each one demonstrates meaningful contributions to their society as well as significant personal growth.

Mircea Eliade: Humanity, Spirit, and Nature

A contemporary of Joseph Campbell, Mircea Eliade (1907–1986) was a Romanian-born historian of religion whose transcultural studies of religious structures and iconography established him as one of the most creative thinkers about religion in the mid-twentieth century. Like Campbell, Eliade was strongly influenced by the writings of Carl Jung, but he had a deep belief in "the sacred" as ontological truth manifested in sacred objects, as contrasted with Campbell's more psychological and metaphorical interpretations of religious belief. But the two men share a common belief in the profound truths of primitive religions and their relevance for secularised, alienated modern man, who lives only in the "profane" world of materialism without experiencing the "sacred" within nature (Cuza 11).

Eliade's most distinctive ideas revolved around the concept of *hierophanies*, spaces and structures that act as portals for sacred energies. His book *The Sacred and the Profane* (1957) established a dichotomy between two modes of being in the world, and two different perspectives on nature. "For religious man," wrote Eliade, "nature is never only 'natural'; it is always fraught with a religious value…It is not simply a sacrality *communicated* by the gods…they manifested the different modalities of the sacred in the very structure of the world and of cosmic phenomena" (Eliade, *Sacred* 116).

Eliade surveys the ubiquity of natural symbolism across the world's religions; these "anthropo-cosmic homologies" serve as "ciphers of various existential situations" (Eliade, *Sacred* 169). In *Patterns of Comparative*

Religion (1963), he devotes each chapter to the way through which natural phenomena act as windows to the sacred in early religions: the sky, the sun, the moon, the waters, the earth, and so on.

For Eliade, the waters symbolise "the reservoir of all the possibilities of existence; they precede every form and support every creation…the symbolism of the waters implies both death and rebirth. Contact with water always brings a regeneration" (Eliade, *Sacred* 130). Mother Earth (Gaia) has given birth to everything, and the human mother is merely her representative on the plane of human fertility. Eliade also notes that "in cases of extreme distress, when everything has been tried in vain…men turn to the supreme being again and entreat him" (Eliade, *Sacred* 126).

Eliade is also known for his interest in the concept of the *Eternal Return*, which sacralises the cyclic patterns of nature in a way that challenges the modern conception of linear time that serves as a foundation for most Western historical and religious thought. This idea, which probably impressed Eliade's contemporaries as rather idiosyncratic, seems more evocative in the light of recent advances in systems theory and ecological science.

The very title of *Elemental* seems to embody Eliade's world view. In the film, the forces of environmental crisis and the spirituality of nature come together within its protagonists as they draw from their personal connections to nature to engage the crises they face. The film begins with a montage overlapping all of its three stories. We see an Indian man dipping himself in the polluted river water, an Alberta tar sands refinery spewing smoke into the air, and an inventor contemplating the ocean. All of the classical four elements—earth, air, fire, and especially water—play prominent parts in the visual panorama of the film. It is humanity's desecration of these elemental forces—the pollution of the waters and the air, the burning of natural resources, and the wounding of the earth—against which these "heroes" are taking a stand.

Rajendra Singh's story focuses on the pollution of water in human society; we see frequent images of dirty, oily water running from pipes directly into the Ganges River. "Without water there is no life," Singh tells a television reporter (00:03:56–00:04:00). Mother Ganges, he insists, is crucial to people's lives and livelihoods, and has been turned into a garbage truck, a danger to the entire region, undermining people's pride, faith, and honour. "The Ganges is India's unity and integrity, like the thread of a garland that ties all of India together. If that breaks, then the garland breaks" (01:24:44–01:24:54). At the film's conclusion, Singh bathes in the clean water of a Himalayan mountain stream and drinks from it in a ritual act of worship.

Eriel Deranger's early memories are of growing up in nature, fishing, and hunting on her reservation, living in sacred space. Eriel's cousin Gitz Deranger comments, "We have been living off this land for I don't know how long now...There was a time, within fifteen years ago, where we could actually just hop in a canoe and paddle down the river. Scoop up water and drink it. That doesn't happen now...There are people that have died...you can come up to my community and I will take you to their graves" (00:58:21–00:58:50).

Referring to her protest march against the tar sands, Deranger states, "There's a whole healing that needs to be done in this walk...there's a lot that needs to be undone, not just environmentally, but spiritually and culturally as well." One of her fellow activists adds, "We're here to heal the land, we're here to heal the water...and the animals and the skies" (01:11:51–01:12:08). The tribal natives with their roots in the land feel the loss of the sacredness of the space and the "profanity" of the industrial exploitation of the land's resources with every day that passes—from the fiery smoke in the air to the poisonous chemicals in the water.

The activism of indigenous peoples is a crucial element in the global environmental movement. Native peoples are often the greatest victims of ecological degradation, as well as the people most grounded in the rhythms of nature. Deranger says, "As native people, we recognise that you do something to Mother Earth and it comes back to us, because we are like the children inside of a mother. We are that connected to the land, because just like an umbilical cord gives him everything he needs, the water and the plants and the air give us everything we need here" (01:26:33–01:26:58).

"We're all creatures of nature," says Jay Harman, "and until the last few thousand years, we were completely immersed in nature...Now more than half the world's population lives in cities, and really have pretty much lost any connection to that kind of experience, of being fully part of nature" (01:03:23–01:04:15). For Harman, nature is sanity, and when we remove ourselves from nature, we lose that connection to sanity.

Pollutants in the atmosphere prevent the air's convection currents from releasing the earth's heat into the ionosphere, keeping the earth from cooling down. His creative inventiveness follows the principles of biomimicry, tracing water's movement in everything from trees to the human body. Although he speaks from a scientific perspective, Harman's fascination with nature has a spiritual tone, noting that every life form on earth starts in a liquid phase. He calls this "nature's blueprint," "what nature uses to make things work and to minimise energies...All of our knowledge as human beings comes from nature...Everything I do is based on this understanding, and this drives me to do everything I can to stop this destruction of the natural world and reconnect us with nature" (00:23:07–00:23:37).

Elemental's three protagonists come from different cultures, different regions, and different spiritual backgrounds. Yet they all have a strong spiritual connection to nature and see its pollution by modern civilisation as "profane," a loss of a connection to the sacred.

Existentialist Themes

For Mircea Eliade, the word "existentialist" was almost itself "profane"; he saw philosophical existentialism as the very incarnation of secular alienation. Yet there is something in the struggles of the "heroes" of *Elemental* that evokes some of the central themes of the existentialist philosophers: the struggle against alienation from modern society, facing the threat of death and/or extinction, the search for one's life purpose or project, and the absurd struggle against impossible odds.

The word *existential* has a dual meaning in the context of *Elemental*. In the general sense, it means, "of, based on, or expressing existence." Anthropogenic climate change is often referred to as an "existential threat" to many of the world's species, and even to the human race. In its philosophical sense, *existential* refers to the themes associated with the writings of certain thinkers loosely affiliated with a philosophical attitude called *existentialism*, among them, Martin Heidegger, Friedrich Nietzsche, Jean-Paul Sartre, and Albert Camus. Both descriptions of *existentialism* play out clearly in the intellectual and emotional lives of *Elemental*'s protagonists. They all face environmental conditions that pose existential threats to their communities, their cultures, and their species. And they must each suffer through a "dark night of the soul" to discover what part they will play in confronting these dangers.

How would an existential approach to *Elemental* compare with the mythological approach? Both viewpoints share a sense of alienation, anxiety, or dread in modern society, which in the case of existentialism emerges not from the repression of the sacred but from its utter absence. To the existentialist, what has been lost in modernity is not direct connection with the sacred but its illusion, leaving man adrift to find his own sense of purpose and meaning, in the face of a death that will bring not an eternal return but a meaningless void. For the existentialist, an authentic life consists not in unity with either society or the sacred, but in absolute freedom and rebellion against the conformity of a society that would extinguish individuality and subjectivity. Jean-Paul Sartre wrote: "[M]an is condemned to be free…because he did not create himself, yet…because once thrown into the world, he is responsible for everything he does" (282).

One interpretation of this rebellion might be that it leads back to nature. Löwith (1952) says that according to Nietzsche, "man must be willing to conform to the cosmic law of the world, which is 'innocent' because it has no aim, no purpose and no meaning. What at first seems to be extreme nihilism…is thus reversed to the highest positivity; to coexist with the natural world of eternal self-creation and self-destruction" (92). Heidegger shared this sentiment; according to Mei (2014), he believed that "the possibility to be through our actions and choices requires in some sense recognizing the primacy of nature. We cannot simply use nature as a resource for our own development. Rather, nature is the source that makes this development possible" (152).

Living an authentic life in the face of the absurdity of death is another central theme in existentialism. Existentialist scholar Walter Kaufmann (1959) wrote, "[T]he acceptance of the fact that I must die…may forcibly remind me of the limited amount of time at my disposal…and thus become a powerful incentive to make the most of my own Being here and now" (83). According to Passmore and Howell (2014), "[N]ature, serving as a symbolic representation of time, serves to increase our existential awareness of our own mortality, that we experience nature as a continuity of the self, and that the human psyche is coextensive with nature… Connecting with nature embeds us more deeply into the existence of life beyond the course of our single lifetime" (381).

Albert Camus' essay *The Myth of Sisyphus* (1955) made the task of continually rolling a boulder up a hill, only to see it roll down the other side, as a metaphor for life. In the context of *Elemental*, no life's task could be more hopeless than the fight to protect the environment from pollution, but the film's protagonists keep pursuing their utopian dreams in spite of the inevitability of failure.

Rajendra Singh's alienation from his technological society and its disconnect with nature runs deep. Science is not his enemy, however, but a valuable ally in diagnosing the pollution of the river. He admits to frustration with India's power structure, asserting that the government never listens to its people, that despite spending millions to clean the Ganges, the river only gets more polluted. In the film, Singh visits a recently constructed dam, where people have been displaced from their land and government promises of resettlement have not materialised. "When they build dams in various places," says Singh, "that kills the river…This dam is a symbol of our destruction. Our entire democracy is run by corporate interests. When a nation is run by corporate interests, then community objectives are left behind" (00:54:40–00:54:56).

Climbing high into the Himalayas, Singh speculates, "If these glaciers melt, the Ganges will dry, die and be destroyed. If this happens, India's livelihood, faith—all will be at stake" (01:18:27–01:18:48). Elsewhere, he confesses sorrowfully, "I feel so ashamed. I don't know why I was born into this country. We are all dead" (00:03:05–00:03:15). He admits that his work consumes him day and night. Even his own son questioned why he worked so hard, but he has now joined his father on his journey. Singh reflects, "During this journey, my heart has been filled with such pain. I feel that in order to heal the Ganges, we need to start a war…I want to become an honest soldier for the Ganges. With strength from the Ganges, we will make the Ganges pure" (01:07:04–01:07:46).

Singh admits to the religious hermit that he lacks the strength to do everything: "Many times, unknowingly, our actions are just God's wish. So when we feel like honoring God's wish, then we find the strength to do it." The hermit replies, "Only if we do it out of love for Mother Ganges will we be successful. If you do it for publicity, nothing you can achieve… Money, name, fame, nothing…it's all going to be dust. Once you leave this world, who are you? What difference does it make?" (01:19:40–01:20:31).

Eriel Deranger represents the survival of family and tribal traditions: "It's been an ongoing struggle for the last eleven years of my life as to whether or not I wanted to have another kid" (00:35:36–00:35:45). However, when her new partner expressed a desire to have a child with her, she agreed. But her teenage daughter started crying when they broke the news to her. Later in the film, Deranger gets to see an ultrasound of her unborn child, to whom she will bequeath the future. Despite her fears, the process of life moves on.

Deranger's alienation from modern Canada has profound roots: "These issues are rooted in deep, deep scars of oppression and colonisation, and we have to begin a process of decolonisation. Part of that process…is giving people their voices back" (01:12:08–01:12:24). Her mother turns the concept of oppression upside down: "White people have been oppressed longer than anybody. They were all tribal. They all were Mother-Earth-based, and something came along and took that away from them first. So they're more oppressed than we are...and they don't even know what to go back to" (01:02:45–01:03:02).

Deranger admits, "I couldn't have a normal job because I wouldn't feel good that I was just letting this happen...I feel bad when I'm not doing something. So for me, it's like do or die" (00:09:01–00:09:44). She has been a protester for so long that losing a paycheck doesn't affect her motivation to continue her activism: "This movement isn't just about my people, my son, my daughter, their futures. It's about the future of Mother Earth, it's about the future of the human species, and all species on this planet" (01:26:58–01:27:13).

Jay Harman admits that when he was young, he was incredibly depressed: "I declared myself bankrupt and I ended up with nothing...I spent a lot of time in the bush, and I thought, there's no place to fall from here. You can't fall when you're in nature." In the world of people and expectations and making a living, you can fail, but "when I came here, there was a complete sense of peace. Nature never, ever lets you down" (01:15:37–01:16:18).

For Harman, "life is about possibilities, so I don't see impossibility... That's just a failure of imagination" (01:24:24–01:24:39). Whenever he sees a problem, he immediately asks, "How does nature do that?" Harman calls himself an optimist; it's his survival mechanism. "I think all I'm reflecting is an observance of optimism in nature, because nature is always optimistic...When it puts out trillions of spores from mushrooms, or millions of eggs or untold numbers of insects, that's optimism" (01:05:24–01:05:39). Global warming may be an out-of-control environmental experiment on a global scale, but Harman has no doubt whatsoever that we can solve the problem: "If we want new technologies, if we want solutions that can really take care of our pressing issues today, they're going to be radical solutions. They're not going to be incremental" (01:05:57–01:06:27).

In the final analysis, regardless of whether the protagonists' motivations derive primarily from spiritual, cultural, or scientific belief systems, the themes of existential philosophy can illuminate their struggles: alienation from modern society, the search for one's life purpose, the absurd struggle against impossible odds, and the threat of death and extinction. They fight their battles and seek their solutions from a place of deep loneliness, in the hope of making a better world for all.

Concluding Reflections

What is ultimately so impressive about *Elemental* is the way it embraces universal spiritual, philosophical, and ecological themes while keeping a sharp focus on individual people in the context of their families, cultures, and personal struggles. In creating the portrayals of its environmental warriors, *Elemental* makes powerful use of the symbolic connections between religion and nature, the mythic structures of the hero's adventure, and the concerns of existential philosophy. These perspectives are richly embodied in the unique personalities and daily struggles of the film's three subjects. Because of the universality of these themes, viewers of the film can more easily identify with these environmental activists, because rather than being portrayed as models of perfection, they are seen as complex human beings, exhibiting the same passions, flaws, and difficult choices that all of us have to make in our own daily lives.

The depth of the ecological crisis can be clearly seen from the complexity of the challenges on both the spiritual and material planes. Reflecting on the roots of religious experience in natural phenomena reminds us how far our contemporary religious belief systems have diverged from their origins. The waste products of our industrial economy come into direct conflict with the resources provided by nature that sustain our very existence. While our economic and political systems seem enmeshed in inertia and corruption, these three crusaders (and many others) persevere daily to raise awareness of the problems and build a critical mass for public protest, institutional change, and technological innovation.

The perspectives of Joseph Campbell, Mircea Eliade, and the existentialist philosophers help provide frameworks for reflecting on our ecological challenges. They offer the opportunity to take the long view, to question the profound spiritual deficits of a culture that allows us to foul our own natural support systems; they challenge us to look within to seek our best selves, and to embrace the necessary struggles rather than retreat from them; and they provide a vista for seeing our own individual lives in the context of the epochal transition that lies before us. Viewers may walk away from this film feeling like they have been put through a wringer—exhausted by the Sisyphusian task of fighting against the inevitable decline of our natural environment—yet inspired by the commitment of those who absorb the slings and arrows of outrageous fortune and keep moving forward, undaunted.

References

Campbell, Joseph. *The Hero with a Thousand Faces.* 2nd ed. Princeton, NJ: Princeton UP, 1968. Print.

Camus, Albert. *The Myth of Sisyphus.* New York, NY: Vintage Books/Random House, 1955. Print.

Cuza, Alexandru Ioan. "Mircia Eliade and Ioan Pertu Culianu: Two Different Paradigms in the Study of Religious Facts." *Philobiblon* 19.1. University of Iasi, Romania (2014): 9–18. Print.

Elemental. Dir. Gayatri Rosham and Emmanuel Vaughan-Lee. Go Project Films, 2012. DVD.

*Elemental*_Discussion_Guide.pdf. Web. http://static1.squarespace.com/static/5480f44ae4b0e96f9a184a94/t/54811a7ce4b0d1770aa0e5ab/1417747068809/*Elemental*_Discussion_Guide.pdf. Accessed from http://www.elemental-thefilm.com/education

Eliade, Mircea. *The Sacred and the Profane.* New York, NY: Harcourt Brace Jovanovich, 1987. Print.

Eliade, Mircea. *Patterns of Comparative Religion.* Cleveland, OH: Meridian Books, 1958. Print.

Ellwood, Robert. *The Politics of Myth: A Study of C.G. Jung, Mircea Eliade, and Joseph Campbell.* Albany, NY: State University of New York, 1999. Print.

Gentile, John S. "The Scholar as Mystic: The Poetic Mysticisms of Joseph Campbell." *Storytelling, Self, Society* 3.3. Wayne State University (Sep.–Dec. 2007): 195–204. Print.

Kaufmann, Walter. "Existentialism and Death." *Chicago Review* 13.2 (1959): 75–93. Web.

Löwith, Karl. "Nature, History and Existentialism." *Social Research* 19.1 (March 1952): 79–94. Print.

Mei, Todd. "The Relevance of an Existential Conception of Nature." *Cosmos and History: The Journal of Natural and Social Philosophy* 10.2 (2014): 138–57. Print.

Paraschivescu, Mihaela. "Joseph Cambell and the Jungian Reading of Myth." *Journal for the Study of Religion and Ideologies* 10.28 (Spring 2011): 216–27. Print.

Passmore, Holli-Anne, and Andrew J. Howell. "Eco-Existential Positive Psychological Psychology: Experiences in Nature, Existential Anxieties, and Well-Being." *Humanistic Psychologist* 42 (2014): 370–88. Print.

Sartre, Jean-Paul. "Existentialism." *A Casebook in Existentialism.* Ed. William W. Spanos. New York, NY: Thomas Y. Crowell Company, 1966. 275–97. Print.

Vogler, Christopher. *The Writer's Journey: Mythic Structure for Storytellers and Screenwriters.* Studio City, CA: Michael Weise Productions, 1992. Print.

The Possibilities of a River and a Ritual Dance: An Ecoethnographic Analysis of Kuttan Aarangottu Vayali's *Bhagavathy aattu*

R. Sreejith Varma and *Swarnalatha Rangarajan*

"Theyyam," the folk-ritual pageant of North Malabar of Kerala, a south-western coastal state of India, which employs a rich tapestry of dance, colour, music, and ritual, has its roots in the fertility cults of agricultural communities that came into existence somewhere after 1800 BC (Varier and Gurukkal 54–55). Richly assimilative in nature, Theyyam or Kaliattam imbibed other cultic practices, including the Puranic stories from the Aryan culture which took root in Kerala from the fifth century onwards. Caste boundaries and taboos are set aside in this versatile art form, whose performative space captures what Arne Naess, the co-founder of the Deep Ecology Movement, calls "the relational, total-field image" (28) and encourages participation from communities across the caste spectrum, from high-caste Brahmins to Muslims, emphasising the importance of "inter-being" in community. The syncretic religious practices of Theyyam advocate an earth-centric religiosity that is not predicated on transcendence. For instance, there are no graven images in stone of the Theyyam deities in the *kavu*, or the ancestral shrine. The gods assume a visible and experiential reality during the Theyyam dance, wherein the abstract is transformed into a palpable living reality when the Theyyam artiste enters the performative space replete with headgear, ornaments, and other paraphernalia.

R. Sreejith Varma (✉) • S. Rangarajan
Department of Humanities and Social Sciences, Indian Institute of Technology-Madras, Chennai, Tamil Nadu, 600036, India

R.K. Alex, S.S. Deborah (eds.), *Ecodocumentaries*,
DOI 10.1057/978-1-137-56224-1_10

The high-pitched, trance-inducing drum beats and the symbolic gesture of looking into the mirror complete the act of crossing the threshold to identify with the deity. The myriads of deities who emerge from the collective unconscious of local history both question hierarchies and bend binaries. T.V. Chandran mentions the emergence of more than 300 Theyyams as a result of this process of assimilation, which resulted in the appearance of subversive deities like Tottunkara Bhagavathy, who questions the sexual crimes of the upper castes; Pottan, who subverts caste hierarchies and questions the concepts of both ritual purity and pollution; and Talakkal Chandu, who revolted against the authority of the British (20). However, Theyyam as a folk ritual and an indicator of folk life is rapidly waning in the Kannur and Kasaragod districts of North Kerala due to the intrusion of a levelling mass culture. Therefore, cognate forms of Theyyam, such as Bhagavathy aattu, have become self-conscious tools of resistance in articulating the importance of the material cosmos from which subordinate social groups, like the Paraya community, draw their sustenance.

The Theyyam dance is a fine example of embedded ecology since its co-ordinates of ritual and myth connect the cosmos with the materiality of the earth and interweaves the themes of fertility and prosperity for the community and its future generations. The possibilities of the river and the ritual performances that flourish on its banks emphasise the matrix of the nature–culture continuum highlighted by thinkers like Latour. In the words of Apfell-Marglin, "culture and society are embedded in nature, and the spiritual is embedded in the material" (qtd. in Chapple 66). The dance, thereby, becomes a way of ordering the cosmos and offers a deep renewal through its ritual enactment of human integration in the world of nature. Charlene Spretnak makes a strong case for similar powerful mythico-ritual projections in which all modern societies are embedded. She observes that these relational fields weave:

> ...a sphere of participation that includes the narrator and listeners (or the reader) along with the forms of universe life that appear in the story. Origins, fruition, waning, and loss—the cycles of arising and passing away—are punctuated with unexpected dramatic turns. All of this is contained in the sacred space unfurled by the telling of the myth. (Spretnak 182)

In a similar vein, Vijaya Nagarajan theorises that the embedded ecologies of ritual act as a social exchange that happens in the interplay between natural and cultural worlds. She writes: "The theory of embedded ecologies characterizes the multiple ways in which culture frames and reveals the complex relationships between nature and culture" (2000: 452). According to Gary Gardner, rituals and embedded ecologies can act as powerful "ecological guardians" and tools for "transforming cultures of consumerism into cultures of sustainability" since they offer a more powerful form of communication than even language (30–31). It is in this sense that Kuttan Aarangottu's film offers the viewer a rich gestalt perception of place and human lives through a seamless interweaving of diverse elements which are rooted in the material countlessness of existence. The river and the ritual dance captured in *Bhagavathy aattu* offer myriad possibilities since they open up the route of ritual communication to express deep, cultural truths about being in and transacting with nature (Fig. 1).

Kuttan Aarangottu's short documentary film *Bhagavathy aattu*[1] is as much an ethnographic study of the eponymous dalit ritual art form it documents as it is a persuasive environmental argument for the protection of the fast-waning Nila, the longest river in Kerala (Leelakrishnan 12; Bhatt and Bhargava 39), on account of a spate of irresponsible anthropogenic interventions in the river. Extensive waste dumping, the mindless mining of river sand, and encroachments on the Nila are posing alarming environmental and health hazards (*Bharathapuzha*, n.p.). D. Padmalal and K. Maya point out, "[T]he major source of revenue…[in the] mineral industry in Kerala state…[is] from the minor mineral sector, especially from sand from alluvial sources" and much of this sand and gravel mined from the river is to meet the growing demands of an accelerated urbanisation (68). In an article titled "Requiem for a River," noted Malayalam writer M.T. Vasudevan Nair offers a poignant description of the current state of the Nila:

> [H]undreds of lorries now queue at every point of access in every panchayat all along the river. Roads are laid right in the middle of the river for quick mining and loading of sand. Thickets have grown into mini-jungles in many places. They shield the gamblers during the day and the illicit distillers at night. (Nair 140)

Such an alarming level of sand extraction, as Ramakrishnan Korakandy points out, "[a]part from causing bank erosion, flooding, salt water incursion, pollution, etc.…also lower[s] the water level and caus[es] drought in the valleys" (1802). It has been noted that the sand mining of the Nila is also crippling Palakkad's agricultural sector. According to a biodiversity study,

Fig. 1. The ritual of Bhagavathy aattu performed in a temple in Kerala. *Source*: Courtesy of Kuttan Aarangottu Vayali

the sea-water incursion into the Nila is specifically debilitating the paddy cultivation in Palakkad, the famed "rice bowl" of Kerala (Deepa A., n.p.). The campaigns organised to help save the mutilated rivers of Kerala, such as the Nila, "from the clutches of the construction lobby" (1802) have, to a certain extent, met their objective of consciousness-raising among the general public. A breakthrough study conducted by the Kerala Water Authority, which took water samples from the Nila, draws attention to the environmental and health problems consequent to the "dump[ing of] garbage and animal wastes from slaughterhouses and discharge[ing of] sewage water directly into [the] river" (Korakandy 1802). The study established that "100 ml of the water [collected from the Nila] contained an average 1,100 coliform bacteria," which makes it unfit for human consumption even though the river remains the major source of drinking water for the vast majority of people living on its banks (*Bharathapuzha*, n.p.). The Nila's ecosystem is also seriously threatened. It is reported that the fish fauna and some bird varieties, such as the stork and egret, are shrinking (Maiti 88). B.K. Sharma, therefore, appropriately calls the Nila "Kerala's catastrophe" (59).

Ironically enough, the jeopardised river Nila, known also as "Bharathappuzha"—where "Bharatha" alludes to both the great Indian epic, the *Mahabharata*, and the country India, as it is called in most regional languages of India, and "puzha" meaning "river"—is the central part of the larger matrix of Kerala's syncretic culture, literature, and ritualistic art forms, such as Theyyam and Bhagavathy aattu, which flourish on its banks. This is the suggestion of the voice-over statement during the opening shots of the documentary—"[the Nila], to Keralites, is not just another river [of the forty-four rivers in Kerala]" (00.00.33–00.00.36)—while the camera slowly pans the currents of the river, flanked by copses of coconut trees, a buffalo wading through its shallow waters, and a crane standing, as if in deep meditation, waiting for a fish to appear, before jump-cutting to a visual of a multi-coloured ceremonial flag atop a temple flagpole, signalling that the temple festival has started.

Bhagavathy aattu: The Ritual

The documentary's primary subject, "Bhagavathy aattu," is a ritual of the dalit community in Kerala called the Paraya,[2] whose traditional occupation is considered to be making *murams*, a type of sieve, usually out of bamboo, and used especially for winnowing grains. It is practised mainly in the villages of North Malabar. The ritual's genealogy can be traced back to an episode in the popular Keralan legend, "Parayi Petta Panthiru Kulam" ["The Twelve Clans born of a Pariah Woman"],[3] where Pakkanar, a Paraya man, was given a boon by Bhagavathy[4]—if she was invoked by erecting a *Pachola panthal* (a green-leaf tent), she would appear and bless him. Thus, the ritual of Bhagavathy aattu typically begins by building the green-leaf tent, made out of coconut fronds weaved together. The "Bhagavathy" is the Goddess of Kodungallur Temple, who is regarded as another "form" of Kannagi, the heroine of the Tamil epic *Silapathikaram*, deified on account of both her chastity and her supreme uxorial devotion to her husband Kovilan. Kannagi's story is recounted to the accompaniment of *cenda* percussions and the clanking of *ilathalam*,[5] and the *nallamma*[6] *devi* is ceremoniously ensconced in the leaf tent. The two male artistes, who will shortly change themselves into gods, start their transformation by first putting on their headgear, which is an invitation to the god(s) to descend into the bodily forms of the artistes.

The first to perform are the deities "Mukkan Chathan" and "Muthappan." The raw materials for their costumes are obtained totally from nature, the chief items being areca nut leaves, the tender leaves of coconut, and the stem of the plant *vanchi*.[7] Peacock feathers are used to designate the beautiful locks of hair.

The colours used in the Bhagavathy aattu—the *trivarna* or tri-colour used to paint the body and face—are indicative of both the community's close contact with nature and an egalitarian outlook that does not segregate mankind in terms of caste or colour. Colours, such as orange, red, yellow, and black, are applied in a neutral manner and do not signify the qualities of evil or good, unlike in elite art forms such as Kathakali, which uses an elaborate colour-coding system: white to signify upper-caste Brahmins and holy men, green for noble warriors and kings, and black for the tribals, lower-caste men, and evil-doers, who represent primitive and uninhibited passions. T.V. Chandran observes that "Teyyam and kathakali thus stand in polar opposition of nature and culture" (66) since Kathakali as an art form was performed in the interior spaces of rich patrons and, therefore, was alienated from the natural world, which is not the case in Theyyam. The dancer who impersonates the fierce Kali in Bhagavathy aattu, daubing his skin with black paint, does not use the colour to indicate social stratification. His sole objective is complete immersion in the deity, whose blackness he tries to emulate using natural colours and dyes.

With a mostly naked upper torso and wearing a skirt with intermittent patterns of red and black, a metallic breastplate, a metal face-mask with a long red strip that resembles a tongue attached to it, bronze anklets around the feet, and a rather elaborate crown, Mukkan Chathan sits on the *peedam* (a stool) and slowly begins to sway to the rhythms of *cenda* and *ilathalam* before standing up to start his dance-like steps in the front yard of the temple. Muthappan, on the other hand, does not have a crown but wears a clanging waist chain and holds a stick in each hand. Almost the entire face of Muthappan is covered with the fibres of a plant called *vanchi*. Between the two, Muthappan is the entertainer, performing tricks to bring smiles to the faces of the audience. They stand on the *peedam* and bless the devotees. During the enactment, Mukkan Chathan, however, also brandishes a curved sword, which, after removing his crown and perhaps still in a trance, he strikes repeatedly against his forehead. The film shows one of the elderly men stopping him from striking himself, by firmly holding his hands (00.05.33–00.05.36).[8] This marks the end of their "performance."

The film then turns its attention to the two other male artistes, who have draped themselves as "Ilaya Bhagavathy" (meaning "the younger goddess") and carry a *marakkuda* (a palm-leaf umbrella). This refers back to the "Parayi Petta Panthirukulam" legend, where Bhagavathy tells Pakkanar that she will only appear holding the palm-leaf umbrella. Both *Thottam* (story-song) and *Attam* (dance) narrate "the birth and growing up of Ilaya Bhagavathy" (00.05.58–00.06.08). Since *Thottam* meticulously narrates the daily routines of Bhagavathy, the artistes who "perform" Bhagavathy have to undertake rigorous *vratas* (penance/regiment). Both artistes wield a sword, similar to that held by Mukkan Chathan, with their dance movements also closely resembling the latter's.

This is followed by a Kali puja. The Kali puja begins by drawing 64 square *kalams*. The different "forms" of Kali that comprise the various avatars of Kodungallur Bhagavathy are also ceremoniously taken to the temple, seeking her blessings.

It is while Ilaya Bhagavathy goes around the village to "personally" meet the devotees and offer them her blessings that the "Mootha Bhagavathy" (the elder goddess) enters the scene. The Mootha Bhagavathy has a total *roudra* (fiery) countenance. The palm-leaf umbrella that she carries, unlike in the case of Ilaya Bhagavathy, is regarded as a symbol of *paricha* (shield). The Bhagavathy aattu ritual formally concludes by replacing the goddess in the green-leaf tent after placating her anger by performing a *guruthy* (a cock-sacrifice). The film shows the Bhagavathy aattu performer biting into the severed neck of the chicken/cock, bobbing his head from side to side as the chicken/cock carcass hangs pendulously from his mouth (00.09.27–00.09.40), rightly reminding one of the ending of Theyyams, Bhagavathy aattu's cognate dalit ritual tradition.

The Embedded Ecologies in Kuttan Aarangottu's *Bhagavathy aattu*

As a cinematic effort, Kuttan Aarangottu's *Bhagavathy aattu* may have its share of flaws. For example, the film is disappointingly short—the entire documentary is about ten minutes long—and as a result, the audience's comprehension of this lesser-known ritual is bound to be limited. But it more than redeems itself by offering an ecocultural slant, placing the dalit ritual within the rich and widely hailed Nila *nadi samskaram* (the Nila river culture). The expository and concluding portions of the documentary invariably strike an ecosensitive note by highlighting the fecundity of the Nila river culture, thereby turning Bhagavathy aattu into one of a broad array of similar cultural specimens.

Rivers, especially in the Indian context, have always been an integral part of the web of relationships where mythology, religion, and culture share equal importance. Both the Indus valley civilisation and Hinduism, for instance, owe their names—and in the case of the former, its very existence—to the river Sindhu, which, as Kley and Reijerkerk rightly point out, is "the Sanskrit word for Indus, the longest running river in the Indian subcontinent" (47). A reverence for rivers and water bodies is one of the defining markers of Hinduism. The rivers Ganges and Yamuna, in particular, have innumerable Puranic references and are deemed the two holiest rivers of India. A mere dip in the Ganges is believed to wash away one's sins and bad karma, acquired over previous lifetimes (Hazen 24). However, it is not uncommon to spot visitors at the Ganges in shock "as they see people using the water to bathe or brush their teeth while at the same time…corpses float past" them (Kley and Reijerkerk 47). The Nila merits comparison with the legendary Ganges and Yamuna since the sacrality of the river is closely associated with Lord Vishnu—the preserver in the Hindu trinity. Thirunavaya, one of the 108 shrines consecrated to Lord Vishnu in the Vaishnavite tradition, is located on its banks. The temple and its precincts are considered to be as holy as Benares, and ritual offerings for the dead—"shraddhas"—are performed on the banks of the Nila, and the popular belief is that the river, like the Ganges, liberates the soul from the endless cycle of birth and death. M.T. Vasudevan Nair, whose birthplace is Kudallur, where the Nila is joined by the river Thootha (Leelakrishnan 25), writes:

> To us, the river was another benevolent Mother Goddess. She discreetly guarded our intimate dreams. Her deep chasms painfully received the frustrations and shame of some of her errant children. The departed dear ones accepted the rituals of our obeisance under her watchful eyes and left peacefully for their heavenly abodes. (Nair 141)

Vijaya Rettakudi Nagarajan, however, usefully reminds us that the attribution of "sacrality" to an object does not necessarily guarantee—at least when temporal restraints (like the morning time for *kolams*[9]) lapse—an "ecological" treatment of the same, where the latter has a suggestion of "conservationism" to it (1998: 279). This is because, as Nagarajan asserts, in the Indian (Hindu) context, there is "an absence of any concept of 'ecology,'" as it is indistinguishably interwoven with other "non-ecological institutions," like culture or religion (1998: 280). She advances her theory of "embedded ecologies" by building on the assumptions of Karl Polanyi, a Hungarian-American economic anthropologist: "The prime reason for the absence of any concept of the economy [in more traditional cultures] is the difficulty of identifying the economic process under conditions where it is embedded in noneconomic institutions" (qtd. in Nagarajan 1998: 280). Analogously, Nagarajan defines "embedded ecologies" as "[e]cological notions, beliefs, and practices [that] are embedded in cultural forms, particularly in religious and aesthetic practices and institutions" (1998: 280). In the final part of this chapter, we seek to demonstrate how the celebrated Nila river culture and Kuttan Aarangottu's own articulation of the same in his documentary film *Bhagavathy aattu* are illustrations of such an "embedded ecology." We will further demonstrate how the documentary *Bhagavathy aattu* also adopts an advocacy function regarding environmental protection during contemporary times when the river and its culture have been shorn of their former glory and face the threat of a slow death.

The Great Nila River Culture

The voice-over narration of the documentary begins with a statement which refers to the Nila as a "maha nadi," a great river (00.00.15–00.00.19). In a fascinating book titled *Nilayude Theerangalilude* [Along the Banks of the Nila], which provides an account of his journeys through the villages situated along the banks of the Nila, the renowned Malayalam poet Alamkod Leelakrishnan echoes this voice-over statement when he notes:

> Rivers have always been the cultural founts of all lands that have great cultures. We know well about the Great River Cultures that grew on the banks of the Nile, the Thames, the Volga, the Euphrates, the Tigris, the Sindhu and the Ganges. In the same manner, the river Nila, aka "Bharathappuzha," is identified by Keralites as a *maha nadi* that has nurtured and developed the unique culture of Kerala. (Leelakrishnan 9)

Of the 44 rivers that form the veritable arteries of the water-rich Kerala, the Nila is remarkable for its sustained influence on Malayalam literature, arts, and various other cultural expressions. The origins of the Nila can be traced to a stream in the "Trimurthy Hill" among the Anamalai Hills in the Western Ghats. According to the legend, the sage Athri meditated at this place (Leelakrishnan 11). Ancient Puranas refer to the Nila as "Pradeepa" and "Perar." The "Bhisma Parva" of the *Mahabharata* also makes allusions to the Nila (Leelakrishnan 13). Leelakrishnan finds three plausible sources for the river's current name "Nila": (1) Since the Nila was the longest river in Kerala, it was called "Neela," meaning "the longest," which, by popular usage, might have become shortened to Nila. (2) When Egyptian merchants carried out trade via Bharathappuzha, they called the river Nila, in memory of the river Nile. (3) Since Bharathappuzha is the "daughter" of Nilgiri Hills, the river was called "Neela" (signifying the colour "blue"), which might have become shortened to Nila. Leelakrishnan, however, considers the third etymological argument to be the most convincing (13).

For generations of Malayalam writers, the Nila has played a central role as a creative muse. Authors such as Thunjathu Ezhuthachan, a sixteenth-century Malayalam bhakti poet who penned *Adhyatma Ramayanam*, *Sri Mahabharatham*, and several short poems like "Harinamakeerthanam," and who is widely regarded as the "Father of the (Modern) Malayalam Language"; Poonthanam, the seventeenth-century bhakti poet and author of *Njanappana* [The Song of Wisdom]; Kunjan Nambiar, the eighteenth-century satirist, "a people's poet" (Menon 327), and the creator of the Thullal art form; Vallathol Narayana Menon, one of the great modern Malayalam poets and the founder of Kerala Kalamandalam, an institution established to promote Kerala art forms such as Kathakali and Mohiniyattam; M.T. Vasudevan Nair, the Jnanpith award-winning novelist, screenwriter, and film director; P. Kunhiraman Nair, who wrote endearing lyrical poems about nature; and Vadakkke Koottala Narayanankutty Nair, commonly known simply as V.K.N., whose novels and short stories are famous for their satirical wit, are but a few prominent ones among the many.

About his deep kinship with the Nila, M.T. Vasudevan Nair famously wrote, "I have often wandered in search of various geographically different places. But, I always come back here. Perhaps it could be called my limitation. Rather than the great seas that hold unknown wonders in their wombs, I like my familiar Nila" (qtd. in Leelakrishnan 26). The river, in almost all his literary and cinematic works, is the Nila. Similarly, while watching the Kura river in the far-flung city of Tbilisi, the award-winning poet O.N.V. Kurup nostalgically declares, "This river is like my Nila in autumn" (qtd. in Kurup xix). "Kuttippuram Palam" [The Kuttippuram Bridge] (1954) by Edasseri Govindan Nair, often referred to as simply "Edassery," is arguably the first literary work in Malayalam that anticipated the Nila's slow death. Standing on the newly built bridge, the poem's speaker observes the plight of the great river that is plodding along beneath the bridge and asks worriedly,

"Mother Perar,[10] will you also turn into a sewage ditch?" Further down the road and resonating with Edassery, P.P. Ramachandran, one of the most promising poets of the current generation, in "Pattambippuzha Manalil" [On the Pattambi River Bank],[11] visualises the river as a dead young girl whose corpse is being dragged away in a lorry. According to Leelakrishnan, the Nila appears as an "emotional presence" in Malayalam literature and the river, to different writers, has taken on different roles—to some she is a friend, while to others she is a lover or a mother (27). The Nila, in her different guises, is also alluded to in a good number of Malayalam film songs, especially those penned by O.N.V Kurup.[12]

Once every 12 years, the 28-day festival of Mamankam, which was also a site of power struggles between the Zamorins of Calicut and the Valluvanad rulers, routinely involving bloody battles, is also held on the banks of the Nila, at Thirunavaya. Numerous families, who specialise in Ayurveda, *yajna* (a Hindu ritual performed to the accompaniment of the chanting of Vedic mantras), and *Kalari* (the traditional martial art of Kerala), still flourish in the villages located along the banks of the Nila. A unique blend of Arabic and Malayalam literary cultures, evident in the popular musical genre of "Mappila Pattu" (literally, "Muslim Song"), originated in Ponnani, where the Nila joins the Lakshadweep Sea. Mappila Pattu, which employs a mixture of poetic metres used in both Arabic and Dravidian literature, has several sub-genres such as "Mala Pattu," "Kappa Pattu," and "Kessu Pattu," to name but a few (Leelakrishnan 33). Folk Hindu/subaltern rituals, such as "Muthappanattam," "Chozhiyum Pilleurm," and "Bhagavathy aattu," also flourish in the riparian areas of the Nila. The Nila river culture, therefore, could be read as a Dravidian variant of "Ganga–Jamuni tehzeeb,"[13] marked indelibly by the strengths of syncretism, synergy, and a tolerance for differences.

The Sacralisation of the Ecological Elements and Its Material Implications

In (Hindu) India, there is hardly any natural resource, be it a river, the earth, or a mountain, that is *not* ascribed personhood. The Ganges thus becomes Ganga *mata* (the Mother Ganges), the earth becomes "Bhu Devi" (the Earth Goddess) and one of the wives of Lord Vishnu, Himalaya is embodied as the masculine "Himavan," and so on, with a mythic tale or reference almost always appended to the name. These "religiously framed ecological conceptions" (Nagarajan 1998: 286) or the "embedded ecologies," though, at times, seem to positively contribute to environmental sensitivity, as in the well-known instance of the Sri Venkatesvara Devasthanam at Tirupati that "actively sponsors...afforestation campaigns," citing the *sloka*: "He that planteth a tree is a servant of God" (Nelson 4). However, this need not always warrant either a sustained ethical action or a sense of care for the environment. According to Nagarajan, this happens because of the presupposition that the resources personified as "goddesses are so large and indeterminant that the waste of humanity is seen as just a speck in the vast expanse of sky, earth, and water" (1998: 277). This is a classic example of ontological humility going awry.

Sacralising natural resources could encourage one to conclude that, owing to their divinity, "they are capable of cleaning [themselves]" (Nagarajan 1998: 277). Furthermore, any human effort to "protect" them could be viewed as human arrogance, "since [humans are] goddess protected, not the other way round" (ibid.). With respect to the Nila, Alamkod Leelakrishnan points out, "We have several [poetic] expressions that metaphorically relate the Nila to the ancient and holy river Ganges" and asserts that one such "expression" calls the Nila "Dakshina Ganga," the Ganges of the South (14). Along the same lines, Rajan Chungath, studying 35 goddess temples ("Kavu") located on the banks of the Nila, addresses the river as "Nila Devi" in his book *Nila Devi Namostuthe*,[14] a title modelled on the goddess chants in Sanskrit. The staggering depletion of the Nila on account of damming, waste dumping, and reckless sand mining begs the crucial question as to whether this deifying of the river has contributed to taking the bounty of the river for granted.

Just as the ecology of the Nila is being religiously defined,[15] Kuttan Aarangottu's *Bhagavathy aattu* also seeks to "culturally remember" (Nagarajan 1998: 272) the river via the *paraya* ritual it documents. Kuttan Aarangottu's attempt, it should be emphasised, is *not* to show the dependence of the ritual on a natural element, but rather to show how a natural element could engender a "cultural text." It conceives the river ecology as "embedded" in the religio-cultural practice (Bhagavathy aattu), which, in turn, is part of the larger meshwork of the Nila river culture and, hence, searches for the same in the last two. The voice-over during the closing shots of the documentary states: "If we don't save the river Nila from dying minute by minute, without its water and beauty ("aishwaryam") we will have to tell our future generations, like a riddle, that a river once flowed along this path" (00.09.42–00.09.56)!

The Nila river culture—the leitmotif in this documentary—is appropriately grounded in the ancient practices of the worship of the Mother Goddess and the other-than-human presences from the natural world associated with her, such as snakes and other totem animals. According to Erich Neumann, all ancient agrarian cultures have emphasised growth and, therefore, the "image of the Great Mother and the sociological matriarchate occupy a central position" (51). The *kalasam* (full vessel), which occupies a central position in the initiation of the Bhagavathy aattu ritual, is a symbol of the plenitude and bountifulness of the earth. T.V. Chandran points out that while invoking the Tayiparadevata (a wrathful manifestation of Kali) in Theyyam, the kalasam is filled with toddy (symbolic of both breast and womb and the life-giving maternal fluids) and decorated with tender areca seeds, which stand for abundance in nature. The Great Mother paradoxically is the life-giver as well as the destroyer/the bringer of death, whose presence is to be feared. Neumann points out that the Great Mother in her terrifying aspect is "the hungry earth which devours its own children and fattens on their corpses...and once fecundated... casting it out again in new birth, hurling it to death, and over and over again to death" (149–50). This terrifying face of the Mother Goddess also makes her the Goddess of War and Hunting.

The documentary depicts a colourful mandala (*kalam*) drawn to invoke the Goddess. In many religious traditions, the mandala is a mystic symbol of the universe often depicted as a circle enclosing a square. Carl Jung considers the "squaring of the circle" to be the archetype of wholeness (qtd. in Lawson 21). The pattern of a circle with a centre is the basic structure of creation and can be seen in the world of nature at both the microscopic and the macroscopic levels. Nature's mandalas can be encountered in diverse forms, ranging from flowers, spiders' webs, and ice crystals to tree rings, sound vibrations made visible, cells, atoms, spiral nebulae, and solar systems (Huyser 3).

Theyyam's mandala art is based on the principles of organicity and the interrelationship of the part to the whole. At the centre of the mandala is the *bindu* (centre-point), which represents both the unfolding of creation into myriad forms and the dissolution of the universe into its primordial elements. By depicting both death and birth as coterminous processes, the mandala depicts a unique organic hologram.

The mandala or kalam in the documentary *Bhagavathy aattu* brings to mind the deep material rootedness of all life in the processes of nature and the interdependent relationships that human beings have with the elemental forces of nature. According to Kuttan Aarangottu, the words "kalam" and "mandala" can be used interchangeably. Hiltebeitel points out that the term kalam, in both Malayalam and Tamil, is polysemous and can indicate a threshing ground and a battlefield, as well as a distinctive altar for animal sacrifices with 16 quadrilateral squares (170). The kalam of the goddess is ritually associated with agricultural bountifulness and fertility, and the rituals associated with the kalam involve the ingestion of this divine plenitude. Hiltebeitel points out that the kalam can indicate "images of the goddess drawn on the ground with powders of five colours (black, white, red, yellow, green) that represent her body as composed of substitutional 'food' offerings…once it is destroyed by being swept up, [it] can be given out as prasadam" (170).

The kalam used in Bhagavathy aattu is a dense mosaic of interlocking triangles drawn in primary colours. The triangle is symbolic of the *yoni*, or the generative aspect of the divine feminine. The three apexes of the triangle stand for the processes of creation, preservation, and destruction, and these aspects of the Mother Goddess as the force of nature (*prakriti*) are seen as an organic triad and not as antithetical principles. The rituals which initiate the mandalic consciousness offer a keen awareness of the principles of organic interrelationships, higher gestalt perceptions, and a deep understanding that "all life is embedded in the fundamental processes of nature. No state is exalted above the other; on the other hand there is the astute awareness of the multidimensionality of life and consciousness and the conceptualizing of the cosmic and individual consciousness in terms of *prakriti*" (Rangarajan 124).

Conclusion

The documentary *Bhagavathy aattu* is anchored in this deep interrelationship between the ecocultural and the spiritual. The felicitously drawn analogy between a Mother Goddess, who roams around the village offering her blessings, and the meandering river Nila, bestowing upon her dependents her natural gifts, cannot be missed. To be sure, the role of the documentarian of *Bhagavathy aattu* is also that of an archivist, recording, in view of the futurity, the existence of a lesser-practised *paraya* ritual and a fast-vanishing river. Leelakrishnan has lamented the fact that the gloomy prognostications of the Nila's future are not a pressing concern for the vast majority of Keralites, a true marker of the decadence of contemporary Keralan culture (123). As if to address this concern, the film *Bhagavathy aattu* employs a self-consciously exhortative tone and is significant for its ethnographical and ecological value. The embedded ecologies of the ritual dance and the river offered by Kuttan Aarangottu's film bring to mind the limits of metaphysics by emphasising our rootedness in the all-embracing gestalts of environment, culture, and other site-specific factors. Embedded ecologies also dispel the notion of human primacy over the natural and perform the important function of highlighting the hubris of unlimited consumption by implying that ontological agency is always constituted by the ecological.

Notes

1. The documentary *Bhagavathy aattu* is an effort by the Vayali Folklore Group, formed in 2004, to, as its website states, "start learning the Traditional/Local Knowledge systems practiced in and around the banks of [the] River Nila (Bharatha Puzha) and share/impart the same to the young[er] generation" (n.p.). "Vayal" in Malayalam means farmland and "Vayali" "stands for *Vayalamma* or the Goddess of Fields in local lingo" (*Vayali's earth songs*, n.p.).
2. The short film fails to explicitly state which community performs the ritual. But since it is indicated that it is practised according to the folk legend of a boon given to Pakkanar, a Paraya man, it has to be assumed that the rituals shown in the film are performed by members of the Paraya community. However, Sasidharan Clari, in his book *Keraleeya Kalanikhandu*, identifies Bhagavathy aattu as a ritual of Nairs (an upper caste), Thiyyas (a backward caste), and Peruvannaan (a caste whose members practise sorcery) (238).
3. The "Parayi Petta Panthiru Kulam" legend goes like this: Vararuchi, an erudite Brahmin and courtier of King Vikramaditya, married a Paraya woman, mistaking her for a Brahmin girl. Since he had overheard some *vana devatas* (forest deities) stating, as he woke up from a nap in a forest, that he would marry a Paraya woman, he took it in his stride, since it had dawned on him that one's fate, after all, was unalterable. But he told the woman that henceforth they would lead a nomadic life, travelling from one place to another. The Paraya woman bore him 12 children. Every time a baby was born, Vararuchi asked her, "Does the baby have a mouth?" When she answered in the affirmative, he would say, "If it has a mouth, God will feed it" and he would ask her to leave the baby where it was delivered. But when she was asked the same question just after her twelfth delivery, the woman, longing to keep the baby, replied, "It has no mouth!" Vararuchi allowed her to keep that baby. But before long, they saw that the baby really did not have a mouth! Vararuchi idolised the baby on the top of a hill and the deity came to be known as "Vayillakkunnilappan" (The Lord without a Mouth). All the 11 babies that they had left behind were adopted by people of 11 different castes and, together, they are referred to as "Parayi Petta Panthiru Kulam."
4. Meaning "goddess" and the feminine form of "Bhagavan," a (male) god. "Bhagavathy," often also spelled "Bhagavathi," is commonly used interchangeably with "Devi," which also means goddess, though Bhagavathy, unlike Devi, carries the connotation of "Mother Goddess" as well.

5. A native version of cymbals, held with both hands and hit against each other.
6. It means "the good mother."
7. Possibly *Rotula aquatica*, which is widely used in Ayurvedic medicines.
8. The temple oracle, "Velichappadu," often also called "Komaram," performs a similar act, mostly while in a trance, and offers the blood from the cut made on his/her forehead to the deity.
9. *Kolam* is an early morning Tamil ritual of drawing patterns with rice flour on the "threshold space" in the front yard of houses.
10. Another name for the Nila.
11. Refers to the stretch of the Nila that flows through Pattambi, a town in Palakkad district.
12. The most well-known among such songs might be "Neeraduvan, Nilayil neeraduvan" (*Nakhakshathangal*, 1986), where the bank of the Nila is imagined as having goosebumps as the (reflection of the) moon "comes to bathe" in the river.
13. The Ganga–Jamuni tehzeeb (the Ganges–Yamuna culture) represents a "composite culture" (*History*, n.p.) characterised by "Hindu–Muslim harmony and friendship," which is invoked through a comparison of the confluence of two of the great rivers of India—the Ganges and the Yamuna (Upadhaya 86).
14. The title of Chungath's book may also be a reference to the poet Vallathol Narayana Menon's (1878–1958) line, "Purana pradhitha nadi, Nila devi, nityam namaste," which can be translated as "Salutations, O Nila Devi, who is well-known ever since the time of the *puranas*," and which appears in the concluding portion of his poem "Bharathappuzha."
15. The religio-cultural embeddedness of the Nila is also evident from its alternative name, "Bharathappuzha." Rajan Chungath asserts, "It was in Thiruvilwamala [one of the five *ghats* of Nila] that Pandavas did *pinda kriya* (offerings to the ancestors). It must be because of the river's connection with the *Mahabharata* that it got the name 'Bharathappuzha.' No other river in India has the same name" (24).

References

A. Deepa. "Mining away the river." *Indiatogether.org*. India Together, 14 June 2005. Web. 9 March 2016. < http://indiatogether.org/sandmine-environment>

Bhagavathy aattu. Dir. Kuttan Aarangottu, Vayali. Film.

"Bharathapuzha nearing death as waste dumping continues." *Haindava Keralam.com*. Haindava Keralam, 7 May 2008. Web. 12 April 2015. <http://www.haindavakeralam.com/HkPage.aspx?PAGEID=6223&SKIN=B>

Bhatt, S.C., and Gopal K. Bhargava, eds. *People of Indian States and Union Territories: Kerala*. Vol. 14. New Delhi: Kalpaz, 2006. Print.

Chandran, T.V. *Ritual as Ideology: Text and Context in Teyyam.* New Delhi: D.K. Printworld, 2006. Print.

Chapple, Christopher Key. "Hinduism and Deep Ecology." *Deep Ecology and World Religions: New Essays on Sacred Grounds.* Eds. David Landis Barnhill and Roger G. Gottlieb. Albany, NY: State University of New York, 2001. 59–76. Print.

Chungath, Rajan. *Nila Devi Namosthuthe.* Kottayam: SPCS, 2013. Print.

Clari, Sasidharan. "Bhagavathy aattu." *Keraleeya Kala Nighandu.* Kozhikkode: Olive, 2012. 238–40. Print.

Gardner, Gary. "Ritual and Taboo as Ecological Guardians." *State of the World 2010: Transforming Cultures from Consumerism to Sustainability.* New York: Worldwatch Institute, 2010. 30–35. Print.

Hazen, Walter. *Inside Hinduism.* Dayton, OH: Lorenz Educational Press, 2003. Print.

Hiltebeitel, Alf. *The Cult of Draupadi. Vol. 2: On Hindu Ritual and the Goddess.* Chicago: U of Chicago, 1991. Print.

"History." *Lucknow Online.com.* Lucknow Online. n.d. Web. 10 March 2016. <http://lucknowonline.com/dukaan/about-2/history/>

Huyser, Anneke. *Mandala Workbook: For Inner Self-Development.* New Delhi: New Age Books, 2007. Print.

Jung, C.G. *Mandala Symbolism.* Princeton, NJ: Princeton UP, 1972. Print.

Korakandy, Ramakrishnan. "State of the Environment in Kerala: What Price the Development Model?" *Economic and Political Weekly* 35.21–22 (2000): 1801–04. *JSTOR.* Web. 12 April 2015.

Kurup, O.N.V. "A Poet's Testament." *This Ancient Lyre: Selected Poems.* Ed. A.J. Thomas. New Delhi: Sahitya Akademi, 2005. xv–xxii. Print.

Lawson, Thomas J. *Carl Jung, Darwin of the Mind.* London: Karnac, 2008. Print.

Leelakrishnan, Alamkod. *Nilayude Theerangaliloode.* Kottayam: DC Books, 2011. Print.

M., Athira. "Vayali's earth songs." *The Hindu.* 28 Dec. 2011. Web. 12 April 2015. <http://www.thehindu.com/features/metroplus/vayalis-earth-songs/article2754876.ece>

Maiti, Moinak. "Economic Analysis of Sand Mining in Bharathapuzha River, Kerala, India." *International Journal of Business Quantitative Economics and Applied Management Research* 1.4 (2014): 85–89. Web. 9 March 2016. <http://ijbemr.com/wp-content/uploads/2014/10/Economic-Analysis-Of-Sand-Mining-In-Bharathapuzha-River-Kerala-India.pdf>

Menon, A. Sreedhara. *A Survey of Kerala History.* 1969. Kottayam: DC Books, 2007. Print.

Menon, Vallathol Narayana. "Bharathappuzha." *Sahitya Manjeri.* 1920. Kottayam: DC Books, 2008. 56–62. Print.

Naess, Arne. *Ecology, Community and Lifestyle: Outline of an Ecosophy.* Trans. David Rothenberg. 1981. Cambridge: Cambridge UP, 2001. Print.

Nagarajan, Vijaya Rettakudi. "The Earth as Goddess Bhu Devi: Toward a Theory of 'Embedded Ecologies' in Folk Hinduism." *Purifying the Earthly Body of God: Religion and Ecology in Hindu India*. Ed. Lance E. Nelson. Albany, NY: State University of New York, 1998. 269–96. Print.

Nagarajan, Vijaya Rettakudi. "Rituals of Embedded Ecologies." *Hinduism and Ecology: The Intersection of Earth, Sky, and Water*. Eds. Christopher Key Chapple and Mary Evelyn Tucker. Harvard: Harvard UP, 2000. 453–68. Print.

Nair, Edassery Govindan. "Kuttippuram Palam." *Edasserykkavithakal*. 1988. Kozhikode: Mathrubhumi Books, 2012. 388–90. Print.

Nair, M.T. Vasudevan. "Requiem for a River." *Waterlines*. Ed. Amita Baviskar. New Delhi: Penguin Books, 2003. 137–41. Print.

Nelson, Lance E. "Introduction." *Purifying the Earthly Body of God: Religion and Ecology in Hindu India*. Ed. Lance E. Nelson. Albany, NY: State University of New York, 1998. 1–10. Print.

Padmalal, D., and K. Maya. *Sand Mining: Environmental Impacts and Selected Case Studies*. New York: Springer, 2014. Print.

Ramachandran, P.P. "Pattambippuzha Manalil." *P.P. Ramachandrante Kavithakal*. Kozhikode: Mathrubhumi Books, 2015. 140–42. Print.

Rangarajan, Swarnalatha. "The Mandalic Consciousness: *Sri Chakra* as Psychocosmogram." *The Trumpter*. 25.1 (2009): 118–32. Web. 23 Feb. 2015. <http://trumpeter.athabascau.ca/index.php/trumpet/article/view/1155/1456>

Schelwald-van der Kley, Lida, and Linda Reijerkerk. *Water: A Way of Life: Sustainable Water Management in a Cultural Context*. Leiden: CRC Press, 2009. Print.

Sharma, B.K. *Environmental Chemistry*. 1994. Meerut: Goel, 2007. Print.

Spretnak, Charlene. *The Resurgence of the Real: Body, Nature and Place in a Hypermodern World*. New York: Routledge, 1999. Print.

Upadhaya, Priyankar. "Communal Peace in India: Lessons from Multicultural Banaras." *Religion and Security in South and Central Asia*. Ed. K. Warikoo. New York: Routledge, 2011. 83–95. Print.

Varier, Raghava, and Rajan Gurukkal, eds. *Kerala Charitram*. Sukapuram: Vallathol Vidyapeedam, 2004. Print.

Vayali.org. Vayali, n.d. Web. 12 April 2015. <http://www.vayali.org/>

Eco-exoticism in *Thorny Land: Invasion of Cheemakaruvel*

Rayson K. Alex, K. Samuel Moses Srinivas, and *S. Susan Deborah*

It is unusual for the creators of a documentary to critically evaluate their own film. Yet we, the authors of this chapter, conceptualised, documented, and edited the film along with P.S. Sachindev and S. Arun Bose. We are undertaking this critical evaluation in order to highlight the sentiments of the people interviewed, which we feel have not been adequately expressed in the film. We subscribe to the ideas presented in this chapter, and agree with many of the people's opinions. However, we have taken efforts to objectively present these ideas so as not to fall victim to any personal bias.

R.K. Alex (✉)
Department of Humanities and Social Sciences, Birla Institute of Technology and Science, Pilani, K.K. Birla Goa Campus, Zuarinagar, Goa 403726, India

K. Samuel Moses Srinivas
Department of English (Self-financed stream), Madras Christian College, Tambaram, Chennai, Tamil Nadu, India

S.S. Deborah
Department of English, MES College of Arts and Commerce, Zuarinagar, Goa 403726, India

R.K. Alex, S.S. Deborah (eds.), *Ecodocumentaries*,
DOI 10.1057/978-1-137-56224-1_11

Literary, Cultural, and Geopolitical Exoticism

In literature, "exoticism" is defined as "the perception and description of difference, or 'otherness'" (Kuehn, "Exoticism in 19th-century literature"). It is a "fascination with literary depiction of the exotic," which is a "complex philosophical, historical and representational issue" (ibid.). The wonder of the exotic in the romantics is evident in Wordsworth's popular poem, "I Wandered Lonely as a Cloud." Floating over the vales and the hills as a cloud would be an exotic experience, and it adds to the mystical allure when the speaker eulogises a crowd of golden daffodils. Although the daffodil is frequently portrayed by artists and poets as a native flower and has been culturally assimilated into the British imagination, like a literary icon, it originally came from Spain and Portugal (Allen, "History of Daffodils and Tulips"). "Where does the daffodil come from?" is an important question in relation to the concept of exoticism. This leads to the understanding that one of the theoretical strands of exoticism is based on cultural imagination, in the sense that exoticism is a vehicle for "pleasure" and "bliss." This cultural exoticism, based on reason and unreason, the normal and the strange, the native and the non-native, creates a new identity and thereby superiority (Till 299). However, exoticism is also a more tangible geopolitical phenomenon.

The root of geographical exoticism is the urge for travel and exploration—physically entering a foreign space simply due to a fascination of the place, people, their culture, and natural products. The "Western 'admiration' for exotic forms reinforces domination of the spaces," which counts as "cultural appropriation" and "commodification" (Welch xvii). Control over the colonies had a specific intention—in Thomas Steel's words, "a certain fascination of feeling," where pleasure could be considered an intrusion (qtd. in Hayes 159). Schmidt considers this colonial intrusion as a European vision of expansion (17). First, Schmidt opines that "exotic geography did much to promote an idea of 'Europe' in the world." Second, he considers exotic geography as the promotion of a "'universal' form of geography" creating an "exotic subject for consumption." Third, exotic geography was "a saleable product"; in fact, it was an agreeable saleable product. And fourth, "exotic geography boast(s) of its capaciousness and abundance" (18). What is abundant is accessible to anyone who "pays" for it. What is abundant can also be easily negotiated. Geographical exoticism pronounces the native–non-native binary of Nicholas Till. It instantly attains a political dimension. The grounds for a political dimension are not merely identifying the exotic elements or the agents of the exotic elements. They are rooted in the effects the exotic has on the native. However, exoticism can be seen as an inherent conflict between two or more entities.

The first part of this chapter will establish the inherent conflict between the native and the exotic, using ecological, cultural, and social perspectives reflected in the 29-minute video documentary, *Thorny Land: Invasion of Cheemakaruvel*, directed by Rayson K. Alex. The second part will question the ethics of such a conflict within the context of a multicultural and multiecological India.

Thorny Land narrates the story of an invasive species of shrub which entered the cultural imagination of the South Indians (especially in Tamil Nadu and Andhra Pradesh) in the late 1870s. The documentary gives an ethnographic account of the uses and harmful effects of the plant through interviewing farmers across Tamil Nadu and some parts of Kerala. The people's perspectives are intertwined with descriptions of the plant by scientists and officials of the forestry department of Tamil Nadu.

Plant Nomenclature as Cultural Exoticism

The vernacular Tamil name of *Prosopis juliflora* is "veelikaattaan"—"veeli," meaning fence, and "kaattaan," meaning one who guards. veelikaattaan is one who guards the fence. To understand this derogatory nomenclature, it is necessary to explain the religious custom of *kaaval deivam*, prevalent in Tamil Nadu. Every village has their protector deities and sub-deities. *kaaval deivam* literally means "the guardian deity." A popular *kaaval deivam* is Ayyanar. Ayyanar, with his 21 or 61 sub-deities like Kaali Amman, Sudalai Maadam, Maari, Madurai Veeran, Ondi Karuppanna Saamy, and others, not only guards the boundaries of the village but also protects the mountains, rivers, ponds, water tanks, and plantations surrounding the village, and also the villagers from any foreign attacks, plagues, floods, and earthquakes (Masilamani-Meyer 58). Thirumaavalavan writes:

> Ayyanar temple is generally located in a corner or at the outskirts of the village; usually the south-west, occasionally in other corners like the north-east also. In some cases the Ayyanar temple is also located on the banks of a river…Villagers consider him [Ayyanar] a powerful deity and believe that at nights he goes around the village mounted on his black steed, accompanied by some of his retinue, all on horseback…Ayyanar commands less

> intense devotion than other personal gods and goddesses because he is responsible for the community as a whole rather than the concerns of the individual. (159)

Although the Ayyanar temples are located on the outskirts of a village, guarding the village from a distance (also a military tactic), they are part of the daily life of the people. Ayyanar is worshipped with fear, devotion, and gratitude. He is offered food, alcohol, and cigarettes. All the deities' "needs" are met for they "stay awake" and guard the village from all ills. The Tamil poetic-spatio-temporal convention of *akam* and *puRam* is particularly useful here. The general understanding of *akam* in conflict with *puRam*, according to Selby, is a "false 'line'" (Selby 33). *akam* is the inside and *puRam* is the outside. The Ayyanar temples could be considered a space in the *puRam*—the space which exists outside *akam*—as *puRam* brings unfamiliar, alien, and conflicting experiences. Contrary to this understanding of *akam* and *puRam*, the Ayyanar temples that exist in the *puRam* space are actually a part of the *akam*—in the sense that they coexist. This *akampuRam*, in philosophical terms, is relational ontology, which is defined as the identity of entities with relationships between them (Wildman 1). Fundamentally, ontology of entities is based on the relationships that they have with each other. In this case, *akam* is an uncontested space because of the existence of *puRam*, at least in the space of memory. However, Ayyanar caters to the needs of the whole village—an holistic approach to *akampuRam*.

The fluidity of the *akam* and the *puRam*, the inside and the outside, brings to light the concurrence of nature and culture. Donna Haraway's popular term *naturecultures* echoes the nature–culture–sacred continuum of Nirmal Selvamony in considering nature and culture as not isolated but a continuity. Ayyanar, who guards the environment, landscapes, people, and community structures, is a holistic entity, a deity. However, the Tamils, by inventing the term veelikaattaan (the fence guard), are strictly "instructing" the exotic plant to "stay outside the fence and guard the fence." It is as if a specific duty is assigned to the plant and nothing more. This is neither *naturecultures* nor continuity of *akam* and *puRam*. It is a clear dichotomy drawn between the plant and the community, the inner space and the outer space, nature and culture—a conscious separation. The name ceemaikaruveel illustrates this dichotomy. ceemai means "that which is foreign." The etymology of the name ceemaikaruveel expounds ceemai as exotic and karuveel as thorny shrub. Tamils distinguished this plant as an exotic native by calling it ceemai. ceemai, in different contexts, could mean "territory," "foreign country," "end," "extremity," "boundary," and "limit," among other things ("Kooththan-cheemaa, Madul-sima, Isan-kani-cheemai"). ceemaikaruveel has an exotic cultural imagination. As aforementioned, as it is a foreign entity, its place is on the boundary or in the territory outside the fence. This is a clear distinction between the native and the exotic. At the end of the film, the farmers, who encounter the plant on a daily basis in various ways, tell the film-makers that they do not want this plant on their land. Devayani from Theni (a district in Tamil Nadu) says, "[W]e don't need it." Devasenan from Puducherry says that even if acid is poured on the plant, only the individual plant will die. The thousands of seeds which the plant has dispersed will "bring the plant to a thousand lives." Further opinions of these adversely affected people will be quoted later, when the ethics of exoticism are discussed.

The Colonial Legacy of veelikaattaan

Brought from Jamaica (Beddome), veelikaattaan was introduced to the then Madras Province, South India, by the British in 1878 (Pasieznik). It was called a royal plant and, thus, has a long colonial history, which gives the plant an historical and political dimension. The plant was introduced into India "owing to its fast growth and drought hardiness" ("Managing *Prosopis juliflora*"):

> It was introduced in India in 1877 (Sind) and, later on, in many parts of India. It was again introduced in 1913 by the then ruler of Jodhpur State in many arid and semiarid parts of Rajasthan, with great success. The State government at the time declared this species as a "Royal Plant" in 1949, and exhorted the public to protect it and encouraged large scale plantations of this species. Later on, it proved to be the most versatile plant for afforestation of the shifting sand dunes, coastal sands, eroded hills and river margins, saline terrains, dry and degraded grasslands and wastelands where rainfall is scanty and erratic. (Muthana, "Prosopis Species Development Programmes in India")

Thampiran from the Agaram village of Chennai, a district of Tamil Nadu, says that veelikaattaan was brought from a foreign place (veLinaadu—"veLi," meaning outside, and "naadu," meaning place) and was sown throughout Tamil Nadu from an aeroplane. Without mentioning a specific date, Thampiran vividly remembers that before the "sowing event," the plant was not present in Tamil Nadu. A website outlines: "[I]t is said that [veelikaattaan] was introduced into Tamil Nadu, South India, by the efforts of the Honorable K. Kamaraj, the then Chief Minister, to boost the livelihood of poor people" ("*Prosopis juliflora*—Fuel Wood Wonder!"). It would have been in the 1950s that Kamaraj introduced the plant into Tamil Nadu to fight droughts and floods. V. Krishnaswamy from Theni, Tamil Nadu, is of the opinion that even though the Chief Minister's intention to sow the seeds in Tamil Nadu was to help the people, it did much damage to the land and the people. The British administration, during the pre-independence era, and the local administration, during the post-independence era, both supported and helped the people in a variety of ways. In doing this, they were not concerned about ecology or the land. The introduction of the plant should also be seen in the context of a lack of awareness, or futuristic thinking, as well as a lack of universal perspective. The above-mentioned reasons for the introduction of the plant, and for the administrators' lack of ecological awareness, are of less importance when compared with the present ecological hazards of the plant. Moreover, our objective is to understand the plant from an ethical perspective. The documentary provides facts and opinions about the wide distribution of the plant, often contradictory, from varying sources, including a government official and local farmers.

Tamil Culture and veelikaattaan: Is the Plant a Coloniser?

The research team travelled across Tamil Nadu and parts of Kerala to document the growth, uses, and harmful effects of the plant. Interviews of farmers and local people, a government official, scientists, and a literary scholar are presented in the documentary in order to give a wide range of socioeconomic and literary views on the plant. Flourishing steadily in Tamil Nadu for over 70 years, veelikaattaan has entered the cultural imagination of the Tamils. This is pointed out in the documentary by Nirmal Selvamony, *tiNai* theorist and literary scholar.

While the narrator, Solano Da Silva, talks about the presence of the plant in folk literature and films, the audience is shown visuals of the popular Tamil film, *Veyil* (2006). veelikaattaan is a constant background image in the film, illustrating the dryness of the place. It is also a prominent feature in the shots where children are shown fighting, accompanied by a song sequence loosely translated as: "[P]laying with the sun, romancing with the sun/fighting with the sun, we played./We roamed around the veelikaattaan jungle where crabs and foxes dwell." The children's playground becomes a "war ground" when the protagonist is attacked in the veelikaattaan jungle. To illustrate how veelikaattaan can heighten the intensity of conflict between people, Nirmal Selvamony narrates the story of a splay-legged woman who is compared to the branches of veelikaattaan in a folk song collected from Sivagiri, a place near Erode district of Tamil Nadu State. The song is about a wife denouncing the extramarital affair of her husband. It is the story of the different ways in which the wife scolds the other woman, using veelikaattaan as a metaphor. The wife says, "[T] he woman who walks with her splay legs, like a veelikaattaan tree, slanders me whenever she sees me." Why is veelikaattaan compared to the other woman? Why does the wife use the imagery of "splay leg"? Does this mean the other woman has a physical deformity? It is unlikely that she is actually splay legged, or has any deformity. Instead, she might even be more beautiful than the wife, in order for her husband to have been attracted by her. If, then, "splay legged" is a behavioural problem, the wife scolds the other woman for being a prostitute, opening her legs wide for any man. The rude derogation aimed at the other woman not only implies her behaviour but also her "otherness." The other woman is not part of the family and the husband's relationship with her is illegal and unethical. This is a good example of the resistance of the *akam* against the non-native *puRam*. However, the Tamil people's understanding of the *akam* and *puRam* did not help in "keeping the plant in its place." It spread and spread widely across Tamil Nadu.

There are various shots of veelikaattaan encompassing large areas of land and of men and women cutting its branches for firewood and charcoal. The abundance of the plant is quite evident in the way it is portrayed in the film. Recently, this plant has gained importance in cultural communities in various parts of Tamil Nadu. There is evidence of people showing resistance towards its widespread growth and of the government's insensitivity towards this. V. Krishnaswamy from Theni describes the plant: "[I]n Theni, the far-flung veelikaattaan has caused an inherent deformity to the native plants; some of the native varieties have shunned its growth due to the heat that veelikaattaan produces." He goes on to say that it possesses some poisonous qualities that can kill human beings. Although in some villages goats are fed with the seeds of the plant, excess intake of the leaves and the seeds can cause premature labour. The narrator in the documentary asks a pertinent question—is the plant a coloniser—and answers it, confirming that it is not a coloniser because most of the plant colonisers integrate with the local culture. The example of *Tamarindus indica* (locally called puLiyamaram in Tamil) is given to substantiate this point.

"Tamarind is a native to tropical Africa and is now widely planted and naturalised everywhere in the tropics. It grows widely in the tropical and subtropical regions of subcontinental and is also planted along avenues, in parks etc" (Kumar et al. 4.18). Although the tamarind is exotic, it has integrated with local Indian plants and is now accepted by the people as one of their own. Tamarind has been assimilated into Tamil culture to the extent that it has a respectful status in the folk songs and literature. It is also surprising to see that the tree has gained a native status among the villagers, so much so that they have almost forgotten that the tree is exotic. The various uses of the tree and its non-invasive characteristics might explain the reasons for this assimilation. In the fifth part of *Pathupaattu* (The Ten Idylls), entitled *Malaipadukadam*, written by the poet Perumkausikanar, one bard encourages another bard to visit the people living in the mountainous regions and enjoy their hospitality. The bard tells the other that he will be offered:

> ...food that will give...[him] sweet liquor aged in bamboo pipes and toddy made from rice, which...[he] can drink without limits. And for...[his] hangover to go, in the morning they will serve...[him] scattered seeds of fruits brought down by waterfalls mixed with sour sweet tamarind fruit and buttermilk in just proportion...At night in the villages with huts you will be served cooked bamboo rice and rice grown on high grounds along with tamarind gravy with avarai [*Cassia auriculata*] beans. ("Pathupaattu")

In a poem entitled "The Scent of Granny" by N. Sukumaran, the poet's nostalgic memories about his grandmother are brought to life through the gastronomic imagery of "banana chips," "tamarind," and "fried fish." The poet recalls: "[E]very time/she steps in/different scents/step in/banana chips/fried in coconut oil/oranges/sweet tamarind/fried fish/granny's call/never enters/the house/with a scentless bag" (Sukumaran). In the first paragraph of an edition of *The UNESCO Courier*, themed "The Living Culture of the Tamils," tamarind is presented as an important ingredient of ethnic dishes prepared in Tamil households and temples:

> The words "Tamil culture" immediately evoke the image of the towering *gopuram* (entrance gateways) of the Hindu temple, at once a commanding grandeur and solemnity; of a beautiful girl, decked out in all her

> finery, graceful and lovely; to the literary minded, of the squatting sage Tiruvalluvar with his palm-leaf and stylus; to the gastronomically inclined, of *idli* (a rice and lentil batter) and *sambar* (lentils, vegetable and tamarind). (Ramakrishnan 3)

"Most of the popular dishes like *sambar*, *morekolumbu* (curds and spices with coconut), *pulikolumbu* (a spicy sour curry with vegetable and tamarind) and the red-hot fish and meat curry are all different combinations of sour and hot tastes" (Bhaskaran 13). The tamarind plant, an introduced species, theoretically belonging to *puRam*, has righteously accomplished the status of *akam*. Considering Schmidt's theory of exoticism, the tamarind plant fails to satisfy his four points. Tamarind, seen as an exotic geographical agent, rather than spreading and taking over, has integrated with surrounding plants and animals and become useful to human communities. The local people also consider it part of their culture, to the degree that it has taken on an "Indian" identity (*indica*). More than simply possessing an economic value, it has entered every Tamil home irrespective of caste and economic status. *Prosopis*, on the other hand, has disturbed *akam* with its invasiveness and has remained as "other."

For the past few years, there has been a dormant activism among the farmers, environmentalists, and common citizens in Tamil Nadu, protesting against the widespread growth of veelikaattaan. The major reason for this resistance movement is the alarming depletion of groundwater levels in various places in Tamil Nadu. *The Hindu,* a popular daily newspaper in India, has published reports about initiatives taken by members of the public to deal with this issue. Town *panchayat* (administrative unit in local areas of India) workers clearing all the veelikaattaan trees from a pond at Kurumbalur, Perambalur District, Tamil Nadu, managed to effectively restore it to harvest rainwater ("Clearing of 'karuvel' pays dividends at Kurumbalur pond"). A mass movement led by the citizens of Poolampadi cleared veelikaattaan shrubs from the Nallathanneer water tank on 25 September 2015 ("Citizens of Poolampadi clear 'karuvel' from water bodies"). Members of the Farmers' Association of Tamil Nadu staged a protest in Trichirapalli, demanding that the State government uproot veelikaattaan trees along the banks of the river Cauvery, and from surrounding water bodies and tanks, to fight the depletion of groundwater levels and to save the standing "samba" crop in Tamil Nadu ("Farmers want Cauvery water, removal of 'karuvel' trees"). Fifty veelikaattaan trees were cut down by the Nainarkulam Municipality to stabilise the water table

("'Karuvel' trees being cut"). Owing to petitions and cases filed against the growth of veelikaattaan, Madras High Court ordered the eradication of veelikaattaan in Tamil Nadu in 2014. Subsequently, the Madurai Corporation banned cultivation of the plant on private properties (Puthiya Thalaimurai TV). Such evidence proves that farmers and common citizens have become aware that the plant is capable of much damage to their land and vegetation. This awareness has led to political engagement exhorting the government to uproot the plant. In the documentary, a few farmers clearly opine that the tree is a menace and should be destroyed. The plant is shown to be in direct conflict with the land and, thus, is considered a most undesirable plant.

The Ethics of Exotic Relationships

An imperative objective of ecocriticism is the preservation of ecological, cultural, and sociological diversities. Ursula K. Heise endorses this idea when she writes, "[D]iversity is valued as a desirable asset in and of itself and reduction of diversity is deplored as ecologically damaging and politically oppressive" (Heise 394). This view, that any kind of diminution of diversity (in cultural, sociological, and ecological terms), be it caused by plant, animal, bird, or human, is at the centre of an ethical enquiry. In the documentary, the Deputy Conservator of Forests of Tamil Nadu, Dr. Archana Kalyani, while describing the plant, maintains that the roots of veelikaattaan can penetrate the earth to a depth of about 53 metres in search of water. It can also adapt to any climatic or soil conditions. The reasons why the farmers and local citizens are protesting against the plant are because it depletes the groundwater level and it is difficult for the plant to be uprooted completely. Contrary to this argument, a few scientists and naturalists consider this plant useful as it is an abode for insects and birds. In the documentary, some are also of the opinion that it provides shade. However, these are not sufficient arguments to establish that the plant is "useful" when it is the cause of so much harm to the land, water systems, and cultural systems of humans and other animals. The plant's ability to change the hydrological system of its habitat should be questioned in terms of ethics. Such arguments cannot be seen just as an anthropocentric approach of material usefulness to humans but, as discussed elsewhere, that the plant adversely affects other animals, as well as the land and water systems. It is within this context that we ask the following question. Is the demand to uproot and destroy the plant an ethical one? Some communities in Ramanathapuram and Thirunelveli, in Tamil Nadu, are wholly dependent on veelikaattaan for charcoal making. Many of the villagers provide firewood to those who prepare charcoal by burning wood. The charcoal is then acquired by the surrounding charcoal companies, which in turn transport it to North India, where it is used as a source of energy and for making specific medicines and *agarbattis*, a kind of incense widely used in Indian households. The preparation of charcoal is a good source of income for the villagers; they earn between Rs.10,000 and Rs.20,000 by preparing 100 sacks of charcoal. One charcoal company official's argument in support of the plant, that the industry supports more than 50–60 families in a village, could be dismissed due to the fact the charcoal-making process has replaced the villagers' traditional occupation due to the changing landscape and seascape, and governmental restrictions on access to the sea. A newspaper report lists the various reasons given for introducing veelikaattaan in the Ramanathapuram District: "The concern in the 1970s and 1980s about deforestation, desertification and fuel wood shortage prompted a wave of projects, leading to the introduction of the species in a large-scale manner" (Jaishankar). Murugan, a charcoal-maker from Ramanathapuram, substantiates this point. He says that he used to cultivate chillies, cotton, rice, pearl millet, and maize on his land, but now he plants more veelikaattaan. It is evident that the short-sighted remedy (encouraging the spread of veelikaattaan) to help alleviate a man-made environmental disaster (deforestation and desertification) has caused another disaster, in a similar fashion.

According to Heise, any agent that poses a threat to the diversity of an ecosystem should be challenged along ethical grounds. Is the government's passivity in responding to this grave situation a remedy? The article titled "*Prosopis juliflora* in the drylands of India: Develop this valuable resource—don't eradicate it" is evidence of this passivity. Referring to veelikaattaan, the article says:

> Invasion of irrigation channels and arable land has affected the agricultural community, and landowners and large, commercial farmers have seen their income threatened. These groups have put pressure on state governments, which have responded by asking forestry departments to stop further planting of *P. juliflora* and begin eradication programmes, notably in Gujarat, Rajasthan, Haryana and Tamil Nadu. However, *Prosopis* is also playing a vital role in sustaining the livelihoods of the rural poor, including the landless, small farmers and artisans—the least vocal groups of society. ("Prosopis juliflora in the drylands of India")

The article lists several ways of using the plant: its timber may be used for furniture, firewood, and charcoal production, and its pods for small-scale pod processing. This shallow management plan does not address the larger issue of the negative impact of the plant on groundwater levels and the soil.

We, the makers of this documentary, take a strong position, calling the plant an exotic, dominant traveller, and an invader, to prove the point that it is an "aggressive ecological imperialist," capable of altering the entire Tamil landscape. The focus of ethics should not rest with the agent(s) but with the activities and the effects of the activities of the agent(s). Here, the agent of exoticism is veelikaattaan. Before the agent's activities leave a permanent and irreparable scar on the landscape, the plant has to be eradicated using all possible efforts—governmental and non-governmental, societal and individual.

It is a complicated issue to discuss ethics in this way, considering the fact that the plant was, in a sense, invited to India, by the people of India, to fight drought. At least in Tamil Nadu, it was consciously introduced by the then Chief Minister of Tamil Nadu, which was a political decision. When the *akam* invites an element from *puRam*, he/she/it is a guest. However, a guest has an obligation to follow the family's social rules as he/she/it has become part of the *akam*. What happens if the guest disregards such etiquette and behaves in a rude and ruthless manner, creating unrest in the family? The guest has become an enemy and is no longer recognised as even being a part of the *puRam* as *puRam* forms a single unit with the *akam*. If the elements of *akam* are aware of the guest's attitude and activities, he/she/it should be removed immediately. veelikaattaan, once a guest, has turned out to be an enemy. It is high time that the *akam* and the *puRam* realises they have an enemy within.

References

Allen, Ray. "History of Daffodils and Tulips." *American Meadows*. Web. 5 April 2016.

Amirthalingam, M. *Sacred Trees of Tamil Nadu*. Chennai: C.P.R. Environmental Education Centre, C.P. Ramaswami Aiyar Foundation, 1998. Print.

Aravind, Singh. "Revegetation of coal-mine spoils using Prosopis Juliflora in Singrauli Coalfield is a harmful practice from an ecological viewpoint." *Current Science* (10 Nov. 2007): 9. Print.

Berhanu, Abiyot. "The impacts of *Prosopis juliflora* on biodiversity in the desert and semi-desert ecosystem, Northeast Ethiopia." Ecosystem Department. Institute of Biodiversity Conservation. Addis Abbaba, Ethiopia. Poster Presentation.

Bhaskaran, Thilaka. "Rice and Ritual: The Tamil art of cooking." *The Unesco Courier*. 37.3 (1984): 11–13. Web.

"Citizens of Poolampadi clear 'Karuvel' from water bodies." *The Hindu*. 25 Sept. 2015: Tamil Nadu. Web.

"Clearing of 'karuvel' pays dividends at Kurumbalur pond." *The Hindu*. 29 Sept. 2015: Tamil Nadu. Web.

"Farmers want Cauvery water, removal of 'karuvel' trees." *The Hindu*. 22 Sept. 2015: Tamil Nadu. Web.

Gadgil, Madhav, and Ramachandra Guha. *The Use and Abuse of Nature*. Omnibus edition. New Delhi: Oxford UP, 2000. Print.

Hayes, Mike. "Consuming Cannibalism: The Body in Australia's Pacific Archive." Ed. Arturo J. Aldama. *Violence and the Body: Race, Gender and the State*. Bloomington and Indianapolis, IN: Indiana UP, 2003. Print.

Heise, Ursula K. "Ecocriticism and Transnational Turn in American Studies." *American Literary History* 20.1–2 (2008): 381–404. Web. 12 April 2016.

Jaishankar, C. "Tree species turning into environmental threat." *The Hindu*. 27 May 2009: Tamil Nadu. Web.

"'Karuvel' trees being cut." *The Hindu*. 3 Sept. 2015: Tamil Nadu. Web.

"Kooththan-cheemaa, Madul-sima, Isan-kani-cheemai." *Tamilnet*. 23 Jan. 2016. Web. 7 April 2016.

Kovel, Joel. *The Enemy of Nature*. New York: Zed Books Ltd., 2007. Print.

Kuehn, Julia. "Exoticism in 19th-century literature." Web. 5 April 2016. http://www.bl.uk/romantics-and-victorians/articles/exoticism-in-19th-century-literature

Kumar, N., Abdul Khader, P. Rangaswami, and I. Irulappan. *Introduction to Species, Plantation Crops, Medicinal and Aromatic Plants*. New Delhi: Oxford & IBH Publishing Co. Pvt. Ltd., 1997. Print.

"Managing *Prosopis juliflora*." Department of International Development. Web. 9 April 2016.

Masilamani-Meyer, Eveline. *Guardians of Tamil Nadu: Folk Deities, Folk Religion, Hindu Themes*. Wieisbaden: Harrassowitz Verlag, 2004. Print.

Muthana, K.D. "*Prosopis* Species Development Programmes in India." *FAO Corporate Document Repository*. Web. 9 April 2016.

Nayagam, Xavier S. *Nature in Ancient Tamil Poetry, Concept and Interpretation*. Tuticorin: Tamil Literature Society, 1953. Print.

Pasiecznik, N.M., P. Felker, P.J.C. Harris, L.N. Harsh, G. Cruz, J.C. Tewari, K. Cadoret, and L.J. Maldonado. *The Prosopis juliflora – Prosopis pallida Complex: A Monograph*. Coventry: HDRA, 2001. Print.

"Pathupaattu – Part V Malaipadukadam." *Peppertrail.com*. 23 Jan. 2015. Web. 11 April 2016.

"Prosopis juliflora – Fuel Wood Wonder!" *Narscimedwonders.blogspot.in*. 23 Jan. 2009. Web. 6 April 2016.

"*Prosopis juliflora* in the drylands of India: Develop this valuable resource – don't eradicate it." *HDRA – The Organic Organisation*. Web. 12 April 2016.

Puthiya Thalaimurai TV. "Court orders removal of Karuvelam trees from waterways." Online video clip. YouTube. 10 Jan. 2014. Web. 12 April 2016.

Ramakrishnan, P.S. *The Cultural Cradle of Biodiversity*. New Delhi: National Book Trust, 2008. Print.

Ramakrishnan, S. "The living culture of the Tamils." *The Unesco Courier*. 37.3 (1984): 3–7. Web.

Rapport, Nigel, and Joanna Overing. *Social and Cultural Anthropology: The Key Concepts*. London: Routledge, 2000. Print.

Schmidt, Benjamin. "Introduction: On the Invention of Exoticism and the Invention of Europe." *Inventing Exoticism: Geography, Globalism and Europe's Early Modern World*. Philadelphia: U of Philadelphia, 2015. Print.

Selby, Martha Ann. "Dialogues of Space, Desire, and Gender in Tamil Cankam Poetry." Ed. Martha Ann Selby and Indira Viswanathan Peterson. *Tamil Geographies: Cultural Constructions of Space and Place in South India*. Albany, NY: State University of New York, 2008. Print.

Sukumaran, N. "The Scent of Granny." *Sangam House*. Web. 11 April 2016.

"The Post-Colonial Hangover." *Foreign Policy: The Post-Colonial Hangover Comments*. 13 Jan. 2012. Web. 30 June 2015.

Thirumaavalavan, Thol. *Uproot Hindutva: The Fiery Voice of the Liberation Panthers*. Trans. Meena Kandasamy. Kolkata: Samaya, 2004. Print.

Thorny Land: Invasion of Cheemakaruvel. Dir. Rayson K. Alex. Ecomedia Team Production, Cockroach in Cocktail, 2014. DVD.

Till, Nicholas. "'An exotic and irrational entertainment:' opera and our others; opera as other." Ed. Nicholas Till. *The Cambridge Companion to Opera Studies*. Cambridge; New York: Cambridge UP, 2012. Print.

Welch, Ellen R. "Introduction." *A Taste for the Foreign: Worldly Knowledge and Literary Pleasure in Early Modern French Literature*. Plymouth: U of Delaware, 2011. Print.

Wildman, Wesley J. "An Introduction to Relational Ontology." *wesleywildman.com*. 15 May 2006. Web. 6 April 216.

List of Contributors

Rayson K. Alex is Assistant Professor at the Department of Humanities and Social Sciences at the Birla Institute of Technology and Sciences, Pilani, K.K. Birla Goa Campus, Goa, West India. He is one of the editors of *Essays in Ecocriticism* (2007) and *Culture and Media: Ecocritical Explorations* (2014). Along with S. Susan Deborah, K. Samuel Moses, and Sachindev P.S., he founded 'The Ecomedia Team' in 2005, which gradually grew into the *tiNai* Ecofilm Festival (www.teff.in). He is one of the technical advisory committee members of the National Council for Science and Technology Communication (NCSTC) of the Ministry of Science and Technology, Government of India, New Delhi.

Divya Anand is an Honorary Research Associate at La Trobe University, Melbourne, and works as Diversity and Academics Program Associate at Novartis Institutes for BioMedical Research, Boston. Her forthcoming monograph, "Re-imagining Nature and Nations: India, Australia and Contemporary Ecopolitical Controversies," explores the connections between modernity and the crises in nature in two different socio-geopolitical locations.

Pat Brereton, Associate Professor and Head of School of Communications, Dublin City University, Ireland, is an environmental communications scholar. His interests are in ecofilm criticism and general environmental communications, as well as in new media studies. He has written monographs including *Hollywood Utopia: Ecology in Contemporary American Cinema* (2005), *Smart Cinema DVD Addons: New Audience*

R. Alex, S.S. Deborah (eds.), *Ecodocumentaries*,
DOI 10.1057/978-1-137-56224-1

Pleasures (2012), and, most recently, *Environmental Ethics and Film* (2016).

Chia-ju Chang is an Associate Professor of Chinese at Brooklyn College, the City University of New York. Her first book in Chinese, *Global Imagination of Ecological Communities: Chinese and Western Ecocritical Praxis* (Jiangsu University Press, 2013), won the 2013 Bureau of Jiangsu Province Journalism and Publication award in China. Her many articles (in both English and Chinese) have been published in the USA, China, and Taiwan. She and Scott Slovic coedited the volume *Taiwanese Ecocriticism: Identity, Environment, and the Arts* (Lexington, June, 2016). Currently she is editing an environmental humanities volume titled *Voices of the Margin: Critical Chinese Environmental Humanities Reader.* She has served as an executive council of the Association for the Study of Literature and the Environment (ASLE). Chang is also the Visiting Kiriyama Professor at the Center for Asia Pacific Studies at the University of San Francisco.

S. Susan Deborah, PhD, is Assistant Professor at the Department of English, MES College of Arts & Commerce, Goa, India. She is one of the editors of *Culture and Media: Ecocritical Explorations*, the first volume in the area of ecocinema in India. Along with Rayson K. Alex, K. Samuel Moses, and Sachindev P.S., she founded a group within OSLE-India, 'The Ecomedia Team,' in 2005, which gradually grew to become the *tiNai* Ecofilm Festival (www.teff.in). She is the recipient (along with Rayson K. Alex) of the ASLE-USA Media Subvention Grant for creating an interactive video space for ecocinema scholarship and for publishing a book on ecodocumentaries (2014). Her interests lie in food studies, ecomedia, and ecocriticism.

Başak Ağın Dönmez, PhD, teaches English at Middle East Technical University, Ankara, Turkey. She is the author of several publications in cultural studies, the posthumanities, and environmental humanities. Ağın Dönmez's chapter, entitled "Bang Your Head and Save the Planet: Gothic Ecocriticism," in *New International Voices in Ecocriticism* (2015) proposed the use of the term "gothic ecocriticism" as an alternative approach to the study of environmental problems through an evaluation of heavy metal lyrics. A further development of the term, to encompass literary studies in environmental contexts, appeared in her journal article "A Gothic Ecocritical Analysis of Bram Stoker's *Dracula*" (2015). Most recently, Ağın Dönmez penned a review essay, "Recent Approaches in the Posthuman Turn: Braidotti, Herbrechter, and Nayar," which appeared in

2016 in a special issue of the journal *Relations: Beyond Anthropocentrism*. Her main research interests include posthumanism, material ecocriticism, critical theory, gothic studies, and cyberculture. She is a member of the Modern Language Association and the European Association for the Study of Literature, Culture and the Environment.

John A Duvall, PhD, is Associate Professor of Communications and Media Studies at the Dominican University of California. He is the author of *Savage Messiah: Ken Russell's Forgotten Masterpiece* and the forthcoming book *The Environmental Documentary: Cinema Activism in the 21st Century*. He earned his MFA in Cinema Production at the University of Southern California and previously worked in motion-picture, sound-effects postproduction in Hollywood, including the Academy Award-winning *Dances with Wolves*.

Simon C. Estok is the recipient of the Shanghai Metropolitan Government "Oriental Scholar" Award (2015–2017) and resident during that period at the Research Center for Comparative Literature and World Literatures at Shanghai Normal University. Estok is also a Senior Fellow and Full Professor at Sungkyunkwan University, where he teaches literary theory, ecocriticism, and Shakespearean literature. Estok's research interests include ecocriticism, early modern European culture, and the relationships between theory and practice. His award-winning book *Ecocriticism and Shakespeare: Reading Ecophobia* appeared in 2011 (reprinted 2014), and he is coeditor of a book entitled *Landscape, Seascape, and the Eco-Spatial Imaginaiton* (Routledge, 2016). Estok also coedited *International Perspectives in Feminist Ecocriticism* (Routledge, 2013) and *East Asian Ecocriticisms* (Macmillan, 2013), and has published extensively on ecocriticism and Shakespeare in such journals as *PMLA*, *Mosaic*, *Configurations*, *English Studies in Canada*, *ISLE (Interdisciplinary Studies in Literature and the Environment)*, and others. His current project is a book entitled *The Ecophobia Hypothesis*, which reviews and expands on work done over the past six years in theorizing ecophobia.

Pietari Kääpä is Lecturer in Media and Communications and Programme Director of the MSc in Media Management at the University of Stirling, Scotland, UK. He is also a Docent in Film and Television Studies at the University of Helsinki, Finland. His work combines transnational and environmental approaches to the study of media and the creative industries. He has published several books and edited collections as well as peer-reviewed articles in different areas of environmental communications. These include the books *Transnational Ecocinemas: Film Culture* *in an Age of Ecological Transformation* (Intellect, 2013, with Tommy Gustafsson) and *Ecology and Contemporary Nordic Cinema* (Bloomsbury, 2014). He currently works on a British Academy project on environmental regulation of Nordic media.

Salma Monani is Associate Professor at the Department of Environmental Studies at Gettysburg College, Pennsylvania. She is coeditor, with Steve Rust and Sean Cubitt, of *Ecocinema Theory and Practice* (Routledge/AFI, 2013) and *Ecomedia: Key Concepts* (Routledge/Earthscan, 2015), and lead editor (with Joni Adamson) of *Ecocriticism and Indigenous Studies* (Routledge/Interdisciplinary Perspectives in Literature, forthcoming 2016). She has also extensively published on explorations of film and environmental justice, film festival studies, and indigenous ecoactivism.

Swarnalatha Rangarajan is Professor of English at the Department of Humanities and Social Sciences, IIT Madras. She was a Fulbright Predoctoral Fellow at Harvard University from 1999 to 2000 and was awarded her PhD from the University of Madras in 2002. She is the first editor of the *Indian Journal of Ecocriticism* (IJE), has served as guest editor for two special issues on Indian ecosophy for *The Trumpeter*, and is on the editorial board of *ISLE.* She was awarded the Charles Wallace Fellowship at CRASSH (Cambridge University) in 2013 to pursue a book project on ecocriticism. She has edited two books entitled *Ecoambiguity, Community, and Development* (2014) and *Ecocriticism of the Global South* (2015) with Scott Slovic and Vidya Sarveswaran, published by Lexington Books. She (along with Sreejith Varma R.) was awarded the ASLE Translation Grant in 2015. Swarnalatha Rangarajan also dabbles in creative writing and her short fiction has appeared in anthologies of publishing houses, such as Penguin, Zubaan, and Westland, to name but a few. Her first novel, *Final Instructions*, which was published by Authorspress this year, has a prominent ecosophical theme.

Nirmal Selvamony is currently Associate Professor and Head-in-charge of the Department of English Studies at the Central University of Tamil Nadu. His major areas of interest are ecocriticism, theory, dramatic theory, and ecomusicology. He introduced the course on ecocriticism into the Indian university system and founded a forum now known as *tiNai* (formerly, OSLE [Organisation for Studies in Literature and Environment]-India) and also a journal (*Indian Journal of Ecocriticism*) to promote this course. Selvamony has also introduced courses on Tamil musicology, and music and literature at the University of Madras. By means of extensive research in *tolkaappiyam* for more than three decades, he has revived the traditional Indian philosophical tradition called *kaaTci* (in a publication titled *tamiz kaaTci neRiyiyal*, 1996 [The Methodology of Tamil Philosophy]), which precedes the *dharshanas*. He has also introduced *tiNai* (ecoregional) musicology, which studies the music of the primordial biomic world regions (see essays published in the reports of Tamil Isaic Cankam, Chennai, 1991, 1995, 1996). For the past 25 years, Selvamony has researched *tiNai* society and published several papers on various aspects of the subject, attempting to consider it as an alternative social order. He has also published ten books and several chapters in books and essays in journals. Moreover, he has taught courses and performed abroad, directed plays, played the guitar professionally, and composed music for plays.

K. Samuel Moses Srinivas, Assistant Professor of English at the Department of English, Madras Christian College, Chennai, South India, is an educationist and an ardent lover of wildlife. His interests are in eco-criticism, postcolonial literature, subaltern studies, Dalit studies, and film studies. He is the coconvener of 'The Ecomedia Team' and is involved in making video documentaries as part of the Team.

R. Sreejith Varma is a PhD student at the Department of Humanities and Social Sciences, Indian Institute of Technology (IIT), Madras, Chennai, India, and is working on contemporary environmental writing in Malayalam. He holds an MA in English and an MPhil in Comparative Literature from the University of Hyderabad. He is a joint winner (along with Swarnalatha Rangarajan) of the 2015 ASLE Translation Grant, which supports their translation of *Mayilamma: Oru Jeevitham*, the life narrative of Mayilamma, the tribal woman leader of the Plachimada anti–Coca-Cola campaign. His translation of two Malayalam short stories by Narayan, the first tribal fiction writer from Kerala, has appeared in the journals *eDhvani* and *Muse India*. His other publications include the article "Writing Back: Narayan's *Kocharethi* as the First Adivasi Novel in Malayalam," in the journal *Luminaire* (2015), and the book chapter titled "The Politics of Land, Water and Toxins: Reading the Life-narratives of Three Women Oikos-carers from Kerala" (coauthored with Swarnalatha Rangarajan) for *Women and Nature?: Beyond Dualism in Gender, Body, and Environment*, edited by Douglas Vakoch and Sam Mickey (forthcoming).

Ecodocumentary Sources

The ecodocumentaries analysed in the essays can be bought/watched from the following sources:

1. *In God's Land* (2012)
   Director: Pankaj Rishi Kumar
   Contact the director at kumartalkies@yahoo.com for copies of the film
2. *Elemental* (2012)
   Directors: Emmanuel Vaughan-Lee and Gayatri Roshan
   Buy a copy of the film from http://www.elementalthefilm.com/buy-now/
3. *Not My Land* (2011)
   Director: Aljona Surzhikova
4. *Timbaktu* (2012)
   Directors: Rintu Thomas and Sushmit Ghosh
   Watch the film at https://www.youtube.com/watch?v=a0Wp-5Mjs7Y or https://www.youtube.com/watch?v=4rl0kc0AkIc
5. *E-Wasteland* (2012)
   Director: David Fedele
   Watch the film at https://www.youtube.com/watch?v=yUCoToorc9M
6. *City's Edge* (2011)
   Directors: Sandeep Kr. Singh, Shweta Radhakrishnan, Gin Khan Siam, Abhishek Yadav and Sharib Ali
   Watch the film at https://www.youtube.com/watch?v=_9fDCHwJ_9E

R. Alex, S.S. Deborah (eds.), *Ecodocumentaries*,
DOI 10.1057/978-1-137-56224-1

7. *The Plastic Cow* (2012)
   Director: Kunal Vohra
   Watch the film at https://www.youtube.com/watch?v=SifRIYqHfcY
8. *Surviving Progress* (2011)
   Director: Mathieu Roy and Harold Crooks
   Buy the film from http://www.amazon.com/Surviving-Progress-Margaret-Atwood/dp/B008D67N8U
9. *The Queen of Trees* (2005)
   Directors: Mark Deeble and Victoria Stone
   Watch the film at https://www.youtube.com/watch?v=xy86ak2fQJM
10. *Bhagavathy aattu (The Dance of Bhagavathi)* (2013)
    Director: Kuttan Arangodu Vayali
    Watch the film at https://www.youtube.com/watch?v=AlAIbi-Gxf8&feature=youtu.be
11. *Thorny Land: Invasion of Cheemakaruvel* (2014)
    Director: Rayson K. Alex
    Watch the film at https://www.youtube.com/watch?v=W01pm7_wUWo

Index

Note: Page numbers with "n" denote notes.

R. Alex, S.S. Deborah (eds.), *Ecodocumentaries*,
DOI 10.1057/978-1-137-56224-1

CPSIA information can be obtained
at www.ICGtesting.com
Printed in the USA
LVHW071452090922
728000LV00010B/707

9 781137 562234